Also by Janna Levin

A Madman Dreams of Turing Machines

How the Universe Got Its Spots

BLACK HOLE BLUES

and Other Songs from Outer Space

BLACK HOLE BLUES

and Other Songs from Outer Space

JANNA LEVIN

Alfred A. Knopf

New York · 2016

THIS IS A BORZOI BOOK
PUBLISHED BY ALFRED A. KNOPF

www.aaknopf.com

Knopf, Borzoi Books, and the colophon are registered trademarks of
Penguin Random House LLC.

Library of Congress Cataloging-in-Publication Data
Names: Levin, Janna, author.
Title: Black hole blues and other songs from outer space /
by Janna Levin.
Description: First edition. | New York : Alfred A. Knopf, 2016. |
"2016" |
Includes bibliographical references and index.
Identifiers: LCCN 20150466924 |
ISBN 978-0-307-95819-8 (hardcover) | ISBN 0-307-95820-4 (ebook) |
Subjects: LCSH: Gravitational waves. | Black holes (Astronomy)
Classification: LCC QC179 .L48 2016 | DDC 539.7/54—dc23
LC record available at http://lccn.loc.gov/2015046692.

Jacket design by Janet Hansen

Manufactured in the United States of America
First Edition

To Warren, Gibson, and Stella

There is nothing more difficult to take in hand, more perilous to conduct, or more uncertain in its success, than to take the lead in the introduction of a new order of things.

—MACHIAVELLI, *The Prince* (1513)

CONTENTS

BLACK HOLE BLUES

and Other Songs from Outer Space

1

When Black Holes Collide

Somewhere in the universe two black holes collide—as heavy as stars, as small as cities, literally black (the complete absence of light) holes (empty hollows). Tethered by gravity, in their final seconds together the black holes course through thousands of revolutions about their eventual point of contact, churning up space and time until they crash and merge into one bigger black hole, an event more powerful than any since the origin of the universe, outputting more than a trillion times the power of a billion Suns. The black holes collide in complete darkness. None of the energy exploding from the collision comes out as light. No telescope will ever see the event.

That profusion of energy emanates from the coalescing holes in a purely gravitational form, as waves in the shape of spacetime, as gravitational waves. An astronaut floating nearby would see nothing. But the space she occupied would ring, deforming her, squeezing then stretching. If close enough, her auditory mechanism could vibrate in response. She would *hear* the wave. In empty darkness, she could hear spacetime ring.

(Barring death by black hole.) Gravitational waves are like sounds without a material medium. When black holes collide, they make a sound.

No human has ever heard the sound of a gravitational wave. No instrument has indisputably recorded one. Traveling from the impact as fast as light to the Earth could take a billion years, and by the time the gravitational wave gets from the black hole collision to this planet, the din of the crash is imperceptibly faint. Fainter than that. Quieter than can be described with conventional superlatives. By the time the gravitational wave gets here, the ringing of space will involve relative changes in distance the width of an atomic nucleus over a stretch comparable to the span of three Earths.

A campaign to record the skies began a half century ago. The Laser Interferometer Gravitational-Wave Observatory (LIGO) is to date the most expensive undertaking ever funded by the National Science Foundation (NSF), an independent federal agency that supports fundamental scientific research. There are two LIGO observatories, one in Hanford, Washington, and the other in Livingston, Louisiana. Each machine frames 4 square kilometers. With integrated costs exceeding a billion dollars and an international collaboration of hundreds of scientists and engineers, LIGO is the culmination of entire careers and decades of technological innovation.

The machines were taken offline over the past few years for an upgrade to their advanced detection capabilities. Everything was replaced but the nothing—the vacuum—one of the experimentalists told me. In the meantime, calculations and computations are under way in groups across the world to leverage predictions of the universe at its noisiest. Theorists take the intervening years to design data algorithms, to build data banks, to devise methods to extract the most from the instruments. Many scientists have invested their lives in the experimental goal to measure "a change in distance comparable to less than a human hair relative to 100 billion times the circumference of the world."

In the hopefully plentiful years that follow a first detection, the aspiration is for Earth-based observatories to record the sounds of cataclysmic astronomical events from many directions and from varied distances. Dead stars collide and old stars explode and the big bang happened. All kinds of high-impact mayhem can ring spacetime. Over the lifetime of the observatories, scientists will reconstruct a clanging discordant score to accompany the silent movie humanity has compiled of the history of the universe from still images of the sky, a series of frozen snapshots captured over the past four hundred years since Galileo first pointed a crude telescope at the Sun.

I follow this monumental experimental attempt to measure subtle shifts in the shape of spacetime in part as a scientist hoping to make a contribution to a monolithic field, in part as a neophyte hoping to understand an unfamiliar machine, in part as a writer hoping to document the first human-procured records of bare black holes. As the global network of gravity observatories nears the final stretch of this race, it gets harder to turn attention away from the promise of discovery, although there are still those who vehemently doubt the prospects for success.

Under the gloom of a controversial beginning and the opposition of powerful scientists, grievous internal battles, and arduous technological dilemmas, LIGO recovered and grew, hitting projections and escalating in capability. Five decades after the experimental ambition began, we are on the eve of the crash of a colossal machine into a wisp of a sound. An idea sparked in the 1960s, a thought experiment, an amusing haiku, is now a thing of metal and glass. Advanced LIGO began to record the skies in the fall 2015, a century after Einstein published his mathematical description of gravitational waves. The instruments should reach optimum sensitivity within a year or two, maybe three. The early generation of machines proved the concept, but still success is never guaranteed. Nature doesn't always comply. The advanced machines will lock on and tolerate adjustments and corrections and calibrations and wait for something extraordinary to happen, while the scientists push aside their doubts and press toward the finish.

As much as this book is a chronicle of gravitational waves—a sonic record of the history of the universe, a soundtrack to match the silent movie—it is a tribute to a quixotic, epic, harrowing experimental endeavor, a tribute to a fool's ambition.

2

High Fidelity

At 6:00 PM the building is quiet for an MIT headquarters. I have to wait outside until a graduate student rolls up and pops off a bicycle to let me in the locked doors, carrying the bike with her up the stairs. "Rai's office is straight down." She points to the hall behind her and wheels away, one foot jumped into the stirrup of the pedal, the other hanging on the same side. She hops off again and is inhaled by a pale office door. Rai's door looks exactly the same and I have the sense it would be easy to mistake offices, like mistaking hotel rooms.

Rainer Weiss waves me in. We skip conventional social openers and speak with familiarity, although this is our first meeting, as though we've known each other for as long as imaginable, the shared experience of our scientific community outweighing a shared hometown or even generation. We lean back in mismatched chairs, our feet propped up on a single stool.

"I started life with one ambition. I wanted to make music easier to hear. As a kid I was in the revolution of high fidelity. Because, look, I was a kid in around 1947. I built hi-fis of the first kind. The immigrants that came to New York, most of them were very eager to listen to classical music.

"See that loudspeaker there? That came from a movie theater in Brooklyn. Behind the screen you had a matrix of those things. I had twenty of them. I lugged them all on the subway. They had a huge fire at the Brooklyn Paramount, and they were getting rid of them. So I had what were movie-studio quality loudspeakers and I had this fantastic circuit that I was building and I had FM radio. And I would invite friends over to listen to the New York Philharmonic and it was unbelievable. You felt like you were in the theater. An unbelievable sound came out of those things."

Rai gestures to the conical metal guts of a circa 1935 speaker. The raw frame has an exaggerated heft that design advances have banished but otherwise looks surprisingly technologically recent, more 1970s indulgence than 1930s necessity. The object fits in visually with the other metal frames from various apparatuses that are stashed around the hive of scientists attending to a gravitational instrument that first imposed itself as a compelling thought experiment in the 1960s. Although he would later find out he wasn't the first, Rai dreamed up a device to record the sound of spacetime ringing. A paragon of scientific ambition, the experiment is now too colossal for this building or even for Cambridge, Massachusetts. An R&D laboratory to develop some of the machines' components is housed in the basement of the building next door, while the fully integrated instruments are constructed on remote sites.

In 2005, Rai molted the venerable role of professor of physics at MIT so he could walk 4 kilometer cement tunnels, affix oscilloscopes to laser beam tubes, appraise 18,000 cubic meters of hard vacuum for leaks, and measure seismic vibrations in dank wasp-infested enclosures. Rai seceded essentially for the privilege to reemerge as a student again but with the elevation of the august title offered the most admired retired—but active—faculty: professor emeritus.

Rai talks with the emphatic rhythms of a generation of New Yorkers, with the quintessentially American phonetics that emerged from an amalgam of European accents. Any German cadence that he contributed to that mix blended away, the familiar timbre reminding me as much of an era as of a region. He was born in Berlin in 1932 to a rebellious father, Frederick Weiss, a communist from a wealthy Jewish family. (Rai's paternal grandmother was from the prominent Rathenau family. "Very German, slightly Jewish" is Rai's characterization.) Rai's mother, Gertrude Lösner, he describes also as a rebel, not Jewish, and an actress. "Somehow they hitched up," Rai says as though there are some things we should never try to understand. "I was the product of that meeting; they were not married yet," he clarifies.

Like every other immigrant listening to the Philharmonic in Rai's living room, he has a narrative for how he got there, a bit of setup to land the tone, but it's not the real action of his life story, which starts soon after the exchange of papers on Ellis Island. Rai's prelude begins in a communist workers' hospital in Berlin where his father was a neurologist. The Nazis infiltrated the infirmary and the district, as they had other neighborhoods. A Nazi plant botched an operation at the hospital, killing the patient and impelling his politicized father to report the incident to the diminished authorities. Like a marauding gang, the Nazis nabbed him off the street in retribution and interned him in a cellar; the family narrative fails to specify where exactly. There he might have rotted—Frederick's own family had disowned him on account of his zealous communism—if he hadn't conceived Rai on New Year's Eve. Rai's pregnant mother and her father, a local bureaucrat in the Weimar Republic, managed his release. Although free to leave, Rai's father was no longer free to stay.

Frederick was pushed over the border into Czechoslovakia. His new family followed him soon after. Rai cannot figure how his parents stopped fighting long enough to conceive his sister Sybille Weiss in 1937. (They used to blame Hitler for their troubled marriage.) As a reprieve from the marital acrimony, the family of four took their first vacation together in the Tatra Mountains on the Polish border. In the hotel lobby, an old wooden Gothic radio with glowing tubes mesmerized Rai during a broadcast of Chamberlain's appeasement foreign policy that would ultimately deliver parts of Czechoslovakia for German annexation. They adjusted the radio dials to lock onto Chamberlain's voice, to register the message without distortion. Rai describes a spooked flock of expatriate Germans, many of them Jews, taking off, trying to get the hell out of the mountains, to get to Prague and then out of Czechoslovakia before this agreement was consummated. "We got out. We were very lucky that way. My father being a doctor is what got him out, because a lot of people didn't."

In New York, his mother supported the family for several years with odd jobs until his father began his own practice as a psychoanalyst. "I went to a school in New York called Columbia Grammar School, which Murray Gell-Mann [Nobel laureate in physics] had gone to. He was several years ahead of me. I was always being compared with him. You know: 'That guy really knew something. You're just a bum.' That kind of thing."

People had frequency modulation radio for the first time, and

Rai knew enough electronics to build an amplifier and augment the quality of the sound. He had a little business going. The first person who bought his system was not an aunt by genetics but by affinity, a woman he called "Aunt Ruth." He doesn't remember what he earned—not that I ask—but he remembers he only charged for the cost of parts. He had become an entrepreneur with a following: the community of immigrants with an appetite for high fidelity. Once they heard the music clarified through Rai's system, the demand grew by word of mouth.

"There were things called shellac records, which were the original records. They had a background hiss. Vinyl records don't have that. They might make a pop. This was a real background hiss. *Shshshshsh*. You see, the stylus was always being driven by the roughness of the surface, and I was trying to think of ways to get rid of that goddamn hiss.

"During the quiet passage of a Beethoven sonata or something like that, when it's slow you always hear the *hish*. And how do you get rid of it? When you have lots of sound it doesn't matter anymore. It gets masked. And I tried to make a circuit that would change the bandwidth of the device as a function of the amplitude of the sound. And I knew that I didn't know enough to do that on my own, and so I knew I wanted to go to college to learn about that.

"I went to MIT for college—I wanted to learn how to do audio engineering well, because that was all I knew. But I very rapidly realized that I didn't want to become an engineer. I switched to physics, and I don't know why. . . . No, I'll tell you; it was really stupid. The Physics Department had fewer requirements than the others, and I was totally undisciplined—I didn't want any requirements."

. . .

Rai assures me everyone on the MIT team is still working. I can see a few shoulders through the open doors. More people are in the laboratory next door. We check into the R&D lab. Experimenters sit on the floor to sift through a bundle of cables or hunch over optics tables, or ratchet some tool or lift their goggles to focus on a bizarrely antiquated oscilloscope used for diagnostics. I swear I see a floppy disk. I promise that the caliber of the technology is mostly impressive, so I kind of gawk at the floppy disk. The physical labor and the meticulous details layer and integrate and feedback and are compounded until a machine is ultimately built. The power structure of the operation is horizontal at some strata. Everybody seems to understand the job, so the collective operates like an elaborate ant colony in constant but not necessarily rapid motion. Without pause, one thing is done and then another. The target of any one scientist's concentration seems incredibly compressed, microscopic given the scale of the thing they're working toward. Everyone is skilled and physically equipped for the awkward pressures on the body and the long hours. A graduate student gingerly shifts a delicate piece on an optics table. Each person contributes to the fabrication of a hypersensitive device that will be ready to record the sounds from space one hundred years—maybe plus a few—after Einstein surmised that spacetime was mutable.

They're constructing a recording device, not a telescope. The instrument—scientific and musical—will, if it succeeds, record Lilliputian modulations in the shape of space. Only the most aggressive motion of great astrophysical masses can ring spacetime enough to register at the detectors. Colliding black holes slosh waves in spacetime, as can colliding neutron stars, pulsars, exploding stars, and as yet unimagined astrophysical spacetime cataracts. The contractions and expansions of spatial distances and of the tempo of clocks move through the universe—through the shape of spacetime—like waves on an ocean. Gravitational waves are not sound waves. But they can be converted to sound through sheer analog technology, much like a wave on the string of an electric guitar can be converted to sound with a conventional amplifier. In a less than perfect analogy, astrophysical calamities are the finger pickers, space-time is the set of strings, and the experimental apparatus is like the body of the guitar. Or, moving up a few dimensions, astro-physical calamities are the mallets, spacetime is the skin of a three-dimensional drum, and the apparatus records the modu-lations in the shape of the drum to play the silent score back to us as sound. Scientists in the control room listen to the detector amplified through store-bought speakers, although they have only ever heard background noise. The *"hish."* *Shshshshsh.*

The MIT facility is invaluable but puny in the larger scheme of the operation. LIGO headquarters is at Caltech, as is another prototype also humbled by the two full-scale instruments at remote sites. Rai asks, "You haven't been to the sites yet? When are you going? Oh, wait till you see it." He leans back in revived amazement. The full-scale instruments are roughly two and a half thousand times longer than Rai's first prototype. I lean back too and consider the proportions. "We don't get so many visitors to the sites."

From the time he started college, his scientific life has been in this mesh of streets in Cambridge, although the moment he stepped out of the subway in Kendall Square he vowed to go back to New York. On that dank morning in September, the industrial corner of the city stank—an unholy mixture of soap made from the renderings of dead animals and their fat mixed in with mayonnaise and pickles. The chocolaty finish was just too much. He didn't go back to New York but toughed his way past the humid fumes on an elongated trajectory that would veer away from Cambridge only for brief but essential intervals. Though no intransigence was suggested during his first few months enrolled at MIT.

"Well, the next thing that happened was that I fell in love with somebody. This was at the height of the Korean War. Like an idiot, I decided that I was just going to take off, and I flunked out. I chased this woman to Chicago. She was a pianist. But she changed my life, by the way. I had never thought of a lot of that stuff, and that's why I started the piano at twenty—or [older], I guess. It was because of her.

"Many years later when I began to think of gravitational waves, I immediately thought, 'Look, LIGO covers the same frequency range as the piano.'

"Anyway, I was totally gaga, crazy in love. I didn't think of what the consequences of that would be. Of course, the girl went off with somebody else. You can never fall in love— I mean, you're not allowed to do that. You know how it is. So I came back. And this was the beginning of my physics. I had such a bad record, having flunked out."

A college dropout in search of work, a forlorn Rai made his way back to MIT and wandered into the Plywood Palace, a ramshackle structure hastily tossed on the perimeter of campus during the emergency efforts of World War II. The life expectancy for the temporary wooden frame was a few years, no more, built to outlast only the war and that by no more than a few months. The creaky, drafty, sooty, uncomfortable but resilient structure survived decades of repurposing, although occasionally a poorly fitted window frame might blow out and down Vassar Street. Building 20 never earned an official name beyond the uninformative numerical system for buildings favored at MIT. No moniker would suit better than Plywood Palace. Though unremarkable in appearance, the Palace was quietly legendary as a half decade of scientists exploited its impermanence. Holes were punched in the plywood walls and ceilings. Pipes were tapped for any resources transmitted overhead or behind thin barriers. Ideas wafted through the three stories along with noise, both boxed in by a hot tar roof and asbestos insulation, as though the very crappiness of the forgiving structure dissolved the inhabitants' inhibitions. At least nine Noble laureates found their prize in Building 20 alongside inspired research into radar, linguistics, neural nets, audio engineering, gravitational physics—a range so resistant to summary that cultural analyses have been devoted to the question, What were the active ingredients that engendered such spirited creativity? After fifty years, defiant in its longevity given the prognosis, there was a wake in 1998 with scientists and neighbors and the children who had grown up in that playground all assembled to watch as the Plywood Palace finally was torn down.

Rai opposed the demolition like the last holdout on the losing side of an eminent-domain battle. The occupants of the Plywood Palace couldn't turn without falling into one another, and the unexpected intersections were priceless and never replicated. Rai once helped a biologist with a dead cat. "Well, a nearly dead cat." The guy's electronics failed while still connected to probes in the pathetic animal. Rai struggled to put his affection for cats aside (he was afraid to look) to help the biologist get some data out of the fading feline. "We were an interesting little community in there," Rai understates.

Sixty years after Rai poked around the shoddy three stories to ask, "Hey, do you need a guy?" he is fundamentally unchanged—though not unevolved. Someone did need a guy, and Rai worked as a lab technician for two years before he made the transition back to student. "I had so much fun as a graduate student. I got married in the middle of all of that, and my wife got pregnant, and that's finally what ended it. I had to get out, okay? But I would have stayed a graduate student forever, because it was fun. I could go from one experiment to the other, and I never thought about money or any of that stuff, so I did one experiment after the other. Some of them were pretty zany." Rai earned a degree and returned to MIT as a professor after stints at Tufts and Princeton. He didn't like the weather in Princeton, he says in way of explanation, swiping away any deeper investigation into his motives.

The idea came to him during a course he taught on the obscure subject of general relativity, Einstein's theory of curved spacetime, as a junior professor. Rai says, "[MIT] figured that, hell, I had been to Princeton, so I must know something about relativity, right? . . . Well, what I knew about relativity you could stick in this finger. I mean general relativity. I'm not talking about special relativity.

"And I couldn't admit I didn't know general relativity. I mean, here I had started this whole research program to study gravity and I tell them that I don't know anything about general relativity. I didn't . . . so okay, I had a major problem on my hands. And I had to be sort of a day ahead of the students. Now all of us have been caught out that way, but I had just been caught out. I couldn't say no.

"So I teach this relativity course. Now, the reason why that figures in the LIGO story is because that's where LIGO got invented, in that course. This was about 1968 or 1969, and I was, as I say, one day ahead of the students. I had a terrible time with the mathematics. And I tried to do everything by making a *Gedankenexperiment* out of it. You see, I was trying to learn it myself. I mean, in the process of learning it, the mathematics was beyond what I really understood. But I kept trying to understand. And the students in the course were very good—I mean, they knew I was bumbling. But at the same time, it was interesting to them, because I would always try to focus on what I knew about the experiments, and that was a rare thing. You see, people didn't teach a course on general relativity and focus on the experiments. . . . They were not walking out. Because I told them a lot of stuff they were not going to get anywhere else.

"The class asked me to discuss gravitational waves. . . . I used Einstein's papers in German because I speak German. . . . The thing I had learned, which was simple and pristine, was that we could send light beams back and forth between things and measure what was going on with them; that was the only thing I really understood in the whole damn theory.

"I gave as a problem, as a *Gedanken* problem, the idea, 'Well, let's measure gravitational waves by sending light beams between things,' because that was something you could solve. The idea was that here was an object. You'd put another object here and make a right triangle of objects, floating freely in a vacuum. And we'd send light beams between them and then be able to figure out, 'What does the gravitational wave do to the time it takes light to go between those things?' It was a very stylized problem, like a haiku, you know? You'd never think that it was of any value."

The idea: Suspend mirrors so they're free to rock parallel to the earth and watch them toss on the passing gravitational wave. Keep track of the distance between them, and their motions will record the changing shape of spacetime. Since light's speed is a constant, the time it takes for light to race the track measures the length of the course. If the light travel time is a little longer, the distance between the mirrors has stretched. If the light travel time is a little shorter, the distance between the mirrors has squeezed.

Precision clocks are not good enough to distinguish minuscule variations in travel time. Rai's idea was to use the floating mirrors to build a far more precise instrument, an interferometer (the roots of the word are "interfere" and "measure"). Instead of bouncing light along one arm, an interferometer sends light down two arms arranged in an L. Laser light is split into two beams, so that one beam travels along one arm of the L and the other travels along the orthogonal arm of the L. Each beam bounces off a mirror at the far end and returns down the respective arm to interfere back at the original apex. The recombined light is then split into two outputs. If the light travels the same distance in each direction, then the light in one output will recombine perfectly so that the output is bright. Light in the other output will combine in perfect cancellation so that the output is dark. If the arms are not the same length the light will come together but imperfectly, out of sync in a sense. The light will interfere with itself. The interferometer is nicknamed ifo, although to my disappointment the colloquial usage for the diminutive is "i.f.o.," as in each letter is pronounced, contrary to the way it's often transcribed without punctuation—"ifo"—as in a smooth sound or word, although the latter still might catch on.

"A lot of people in the class got captivated by it.

"What I got out of that course were graduate students. We would have evening sessions—it was a wonderful lab—I kept thinking about this crazy business with the floating objects and the light traveling between them. It looked like it wasn't nuts to do that."

After a summer stewing over the idea, influenced by theoretical progress and ongoing experiments in his lab, Rai built a small prototype in the still-standing Plywood Palace. The little instrument with mirrors at the apex and far ends of a 1.5 meter L was not sensitive enough to detect any veritable changes in the shape of spacetime. But it was a proof of concept and focused their intentions so that Rai and his first students devised algorithms to study the hypothetical data in the event an exploding star sent a burst of gravitational waves to the Earth or an orbiting pair of black holes rang spacetime, escalating in pitch until the pair collided into one silent big black hole. They managed to keep the "goddamn thing" running but had to work at night after the subway closed since the entire place would shake every time the Red Line rattled past MIT, sending the instrument's mirrors swinging uselessly. Rai managed to close off Vassar Street for a weekend. The thing would fall out of alignment every time a truck took that back route. He straightens, cheeks high like hot-air balloons strung to the corners of his closed smile as he describes the feat—a functioning prototype under such absurd conditions, although on reflection the absurd conditions might have been just what they needed.

The hasty construction of the Palace reflected an unpreparedness the government intended to amend in the wake of World War II. Roughly shaken out of introversion, the country did not have an army of trained scientists and engineers, and this deficit was believed to hamper military research. Under the pressures of the war, incited by that urgency, technologies were constructed as suddenly as the building, if with higher production value. The tense motivations produced some of the most crucial technological advances during the war—think radar and microwave engineering—and those were quickly integrated into the quotidian concerns of life during peacetime. Although in the 1960s the primary lab in the Plywood Palace still survived on a grant from the joint services, Rai assures me that the support came with no strings or directives from the military except that the money be used to train scientists and engineers in pursuit of interesting research.

"No, no, the work wasn't classified. The military was absolutely the most wonderful way to get money. Their mission at that time—and that's something that's grossly misunderstood by all the people that got into trouble with Vietnam and everything else—the military was in the business of training scientists. They wanted not to get caught again the next time there was a need for a Manhattan Project or a Radiation Lab . . . and all they wanted to do was train good scientists, they didn't give a goddamn what they were going to work on."

Building 20 was a proof of concept, practically a shrine of productivity full of industrious civilians born in the land of originality and freedom and all of that rhetoric. Less tense, possibly more joyous exploration rode the momentum of the brash successes of the war effort and continued for the five decades of the Palace. Another legacy of the war was a system of funding for that exploration. Rai considered the freedom that the military support allowed to be a major attraction on his return to MIT as a professor. "You didn't write a proposal; I mean, you went to the head of the lab and you asked. So they gave me fifty thousand dollars, which was a huge amount of money. They scraped it together from someplace, and I bought a lot of stuff to build the 1.5 meter prototype."

In the quirky climate of the Plywood Palace, the notorious academic pressure to publish or perish also abated, so that Rai could adhere to simple principles and high standards. No incomplete results, no unfulfilled ideas or shoddy experiments should find their way into refereed journals. There's an element of academic social climbing in rampant publishing that Rai shunned. "One of the things about me is that I never published very much and that has caught me up many times. So I don't know, in the end it may have caught me but good. . . . Well, later it cost me plenty."

Rai was bold and practical and effective but not politically ambitious. He pursued experiments out of raw curiosity, indifferent to his career trajectory. He says, "I didn't even think of the tenure clock. It was not in my consciousness. I was a professor, they had just hired me, and I was going to try to do the most interesting thing I could think of. The hell with it." His unfettered attitude let him explore and take risks. His unfettered attitude also led him out of the comfort of the mainstream. Astrophysical sources for gravitational waves were poorly understood. Not only was the experiment exacting, with an uncertain future, a slow burner with unknown boiling point, but there might be nothing out there to make it play. Even if he succeeded, he could fail.

"I was being told by people in the department that they were beginning to worry about what the hell was going to happen to me. They figured that this program that I had started was so long-range that maybe I should do something with more immediate results. And I'm not the type of guy who takes advice like that, okay? Because I'm going to work on a problem that's important; I don't give a goddamn how long it takes.

"Bernie Burke was the head of the astrophysics division and became my mentor. I didn't want Bernie as my mentor, okay? But he imposed himself on me as my mentor. That's Bernie's style. And he was trying to give me advice. He said, 'Look, you're not going to ever get tenure'—and I didn't know what tenure was—'if you continue this way, because none of the things you're doing are of any significance, really. And you haven't published anything—not enough,' and all that sort of stuff. 'You've got to do something and get published.'"

He couldn't keep a student on the ifo for long. There was too much technology to develop in time for a degree. The lifetime of the project would exceed the time for a graduate degree by many multiples, although Rai hadn't yet projected by how many. He also had come to accept that his colleagues would sneer at the whole idea. A fully operational machine was off in a far future. He had no defense to the concern repeatedly vocalized: Maybe no astrophysical phenomena are calamitous enough to ring space and time loud enough.

Here Rai came to a proverbial fork in the road. To achieve scientific goals, the instrument needed to be big. Very, very big. A few thousand times bigger than his prototype, a few kilometers at the least. Longer than the MIT campus. The absurdity of the scale increase could be compelling enough reason to quit. He wasn't publishing. His students had to be moved to other more mainstream projects. Rai could have been denied tenure, which is equivalent to getting sacked. And the comfort of an exploratory lab funded with military support from the joint services ended fairly suddenly. "This was completely corrupted by the Vietnam War, okay? . . . Unfortunately the Vietnam War interceded and the Mansfield Amendment came along and that's what did me in. . . . That was the beginning of the end of military support. Somehow people had gotten the idea that the scientists were then indentured to the military. And that was a very bad thing, because, you see, they were so angry about the Vietnam War. It was part of the anti–Vietnam War movement . . . the stuff I was working on was irrelevant to the military. So I immediately, for the first time in my life, wrote a proposal."

Rai wrote a proposal, this would have been around 1973, to the National Science Foundation to continue the work on the 1.5 meter prototype. His proposal was declined. Without funding and without a fair plan to keep students in the lab, Rai redirected his energy to a different cosmological experiment, measuring the remnant glow from the big bang. (Here he was grateful to Bernie Burke for his interfering and subsequent good advice, which steered Rai and his students to important cosmological experiments as an escape.) He managed, thrived even, but his idea, which just might not be nuts, seemed doomed.

Roughly a year after the disappointment of his rejected proposal, Rai received a phone call from a German physicist at the Max Planck Institute, "a guy, Heinz Billings. He wants to know how far this interferometer had gotten . . . they were fetching around looking for the next step, and they were really turned on by this idea." Rai could not figure out how Billings knew about his little ifo in Building 20. The only publication he had submitted on the work was an internal report, which might have some modest distribution but would not be represented in your typical library. When he pressed the question, Billings admitted that he learned about the work from Rai's unsuccessful NSF proposal. Rai suspects that all the serious experimentalists in gravitational-wave research would have been sent his proposal, their opinions on its merits solicited by the NSF.

"At the time we hadn't gotten all the way to where [the ifo] was functioning. What happened, however, is that they started working on it. I mean, you can't stop people; you can't do that. And the Max Planck group in fact did most of the early development, because they had the money. I was always very jealous of that. They had the money, and they had a large group of very experienced professionals. . . . And they went immediately into interferometers—this was about 1974, probably—and I couldn't go forward."

Rai was glad the Germans were progressing. He was jealous too. He complained to the NSF that his rejected proposal was endorsed in Germany in the most meaningful way any scientific proposal could be endorsed, and his well-founded gripe impelled the NSF to give him some money, enough to finish the MIT prototype. Meanwhile, the Germans were very well organized engineers with access to funds, "and they did a spectacular job building this thing." The German ifo was 3 meters long and beautiful, but, like Rai's, too small to detect any gravitational waves. It was a toy, the stylized miniature car of ifos.

The idea had spread and was becoming something, a physical thing, growing in scale and technology. The idea was in other scientists' hands, literally their hands, soldering and welding and bolting, dragged out of an atmosphere of ideas into real metal and laser light. Rai's disadvantage was significant and, he understood, essentially insurmountable. He couldn't build the real thing, the full-scale machine, the ultimate recording device, the insane astronomical pinnacle of audio engineering. He would watch while others made his haiku physical. He kept at it still, developing the instrumentation on the side, swapping students in and out of the ifo lab while succeeding on other experimental fronts. Rai began life with one ambition, high fidelity—"to make music easier to hear"—and that ambition was tied to this far-out, underrated project in a ramshackle lab that could never compete.

Rai says, "Then I met Kip. That's the next big event."

3

Natural Resources

Kip Thorne is an iconic astrophysicist, a brilliant and influential relativist. He wears an inverted white beard, balanced apex down, framed by darker whiskers like a white shirtfront glowing against chestnut lapels. His long hair is long gone but the bohemian spirit of the '60s and '70s is ineradicable. So few astrophysicists are as influential as Kip that the acclaim is almost an eccentricity. His specifics—like hair placement, length, and color—gain undue attention, eccentric under high magnification.

In the late 1970s, already an accomplished professor at Caltech, Kip wanted to get into something big. Though Kip is a theorist, bearer of a profound and careful intellect, able to roam the vast ranges of the severely abstract, he wanted Caltech to get into something observational, something real. With prerogative and talent as cargo, under obligation to a comprehensible universe, he wandered unfamiliar streets on a trip to the Northeast in the hopes that a brisk constitutional could dislodge an answer to the question, "What am I supposed to be doing with all that I have?" He might not have literally looked up into the sky like a prospector assessing a valuable natural resource, but he did wonder which of the universe's assets he should siphon down to Earth. He decided, though it might have been more of a realization, that he wanted to bring Caltech into the pursuit of gravitational-wave detection.

Kip Thorne's family moved to Utah before the railroads were built. Traditional Mormons for generations, his educated parents were, in that context unconventionally, feminists. His father, D. Wynne Thorne, a soil chemist, was a professor at Utah State University. Nepotism laws of the time precluded his mother, Alison (Cornish) Thorne, a PhD in economics, from holding a professorship at the same university as her husband. Although she did start a women's studies program there, she did so with no official position. Long after his father passed, his mother sat the family down, three sisters bookended by two brothers ("A small Mormon family," Kip quips), and announced they should break from the church in response to their practices concerning women. The church happily excommunicated the girls, but not the boys. "We had a harder time convincing them," he laughs. The local paper ran a headline banner on the front page for his mother's obituary: "Old Radical Dies." Kip's admiration for his mother feels fresh, and I suspect that his free spirit—a phrase I could have made up just to describe Kip—is hereditary.

Kip aspired to drive a snowplow, but his career redirected at the age of eight when his mother took him to an astronomy lecture. The introduction could not have been so fortuitous. As though too well inscribed with a mathematical ability nurtured under Utah's firmament, Kip seemed destined for astrophysics. By the time he met his influential mentor John

Archibald Wheeler, no dreams of snowplows complicated his determination.

His famed advisor Wheeler taught the first Princeton course on relativity in 1952, about a decade before Kip enrolled. Wheeler's intention was as much to learn the subject as to teach it, apparently a standard tactic for physics professors. The new calling of general relativity would last the rest of Wheeler's life. He mentored forty-six PhDs in physics (difficult not to mention his most famous student, Richard Feynman). He is known as the grandfather of American relativity, producing the first wave of the great American relativists, among them Kip, and many of the subsequent waves as well. I remember seeing him at the notorious Princeton lunches, where visitors are expected to present their research to the table. Wheeler was royalty, in his eighties by then, straining to hear with the help of an ear trumpet. (Did I imagine the ear trumpet?)

Wheeler turned to relativity as he emerged from the nuclear weapons program. He helped to design and to use the plutonium production reactor in 1942 until the end of the war. The plutonium plants were enormous, devised to produce 250 million watts of power, not quite twice the power required to illuminate a liberally outlined Times Square. The raw electrical power was invested in a device, carried in a fighter plane over a target, and freed to sail down to Earth, detonating the mortifying equivalent of 20 kilotons of TNT. The plutonium fission bomb lit up an American desert and inspired Oppenheimer's memorable translation of the Bhagavad Gita, "Now I am become Death, the destroyer of worlds." Within a month, a uranium fission bomb Little Boy was detonated over Hiroshima, and three days later a plutonium fission bomb Fat Man was detonated over Nagasaki.

Convinced of his civic duty, Wheeler joined the war effort despite the personal sacrifices, the strain on his family, the suspension of his own scientific pursuits. Before he felt called to service, he would listen to the radio in the Fine Hall tea room at Princeton, surrounded by the overly civilized airs of the British universities emulated, and despite his friendships with the intellectual émigrés, including his close friendship with Albert Einstein, he found the rumors of German atrocities impossible to believe. He didn't believe them. By his own description, his colleagues were consternated when they came across him casually perusing the propaganda that came through his membership in the German Physical Society. In his autobiography, Wheeler describes his sympathy for the German state, his conviction that German dominance would translate to European stability, his parents' chastisement, and the gradual ebbing of his sympathy for Germany as the war progressed. He writes frankly about his error in judgment, which he fully accepted, along with his parents' assessment, as news of atrocities accumulated. He admits, "It is hard now, more than fifty years later, to recapture my frame of mind at that time. . . . Even when I was doing everything I could to help defeat Germany, I clung to the belief that people are fundamentally decent everywhere. . . . By the end of the war I knew better. But not until I visited Auschwitz in 1947 was the full horror of German barbarism brought home to me."

Wheeler determined to contribute to the effort as the United States declared war on Japan on December 8, 1941, the day after the attack on Pearl Harbor. Physicists were drained from academia and flowed around the country to find application of their skills in the Plywood Palace at MIT, in nuclear research facilities in Los Alamos, New Mexico, and Oak Ridge, Tennessee. By early 1942, Wheeler was in service in Chicago, then

Delaware, before facing the giant plutonium reactors in Hanford, Washington, in 1944, dedicated to providing the United States with the atomic bomb to defeat Germany. Within a few weeks, the reactors powered up, Wheeler received news that his younger brother Joe, deployed in Europe, was missing in action. The urgency that already motivated Wheeler in the struggle to produce a nuclear weapon intensified. He writes, "For eighteen months, until it was discovered in April 1946, Joe's body, disintegrating to bones, lay with that of a buddy in a foxhole on the hill where he was killed." When challenged on the use of the atomic bomb, he would respond as he did in his autobiography: "One cannot escape the conclusion that an atomic bomb program started a year earlier and concluded a year sooner would have spared 15 million lives, my brother Joe's among them."

In 1950 Wheeler joined the effort to build a hydrogen bomb out of concern for national security, given escalations in the Cold War. Many of his friends and colleagues rejected his reasons and criticized his involvement. He was pained by the divisiveness but unapologetic. Even Oppenheimer initially opposed the program to build a hydrogen fusion bomb, a weapon of potentially unbounded power. (Oppenheimer would later back the project.) Although Wheeler did not testify in the hearing in 1954 that would strip Oppenheimer of his security clearance (the notorious Edward Teller did), Wheeler was not entirely unsympathetic to the testimony or the decision. I'm using convoluted double negatives because I don't feel myself in a position to more precisely define Wheeler's state of mind, but Kip does. Kip informed me, based on his own discussions with Wheeler, that I could land the grammatically simpler phrasing: He was sympathetic.

Wheeler also was not entirely against the House Un-American Activities Committee. He was not entirely against the anticommunist fervor that purged academics from their ivory-tower ranks for crimes of silence either. (Rai's father, by the way, had plenty to fear during those censorial times. He destroyed photos of himself with Lenin and Trotsky and "lied like a rug," according to Rai. He enlisted Rai to transcribe his patients' files into a code based on the Greek alphabet (α for a, β for b, etc.), an attempt to conceal any references to communism, which had been a sort of fashion among Europeans—communism that is, not the code. Anyone suspected of communist activity would be pressured to give up names, including the name Frederick Weiss. And maybe Wheeler wouldn't have been entirely against this sort of thing either.)

Wheeler was able to direct his attention back to pure science after he felt the urgency of his call to national service subside some. But his experience with nuclear power actively molded his scientific interests. The hard-earned knowledge of nuclear physics had led to horrifying new ways to kill people. Immune to moral tone, the same dispassionate physical laws operated off the Earth. The hard-earned knowledge of those laws also led to grand resolutions of venerable questions, like, How does the Sun shine? Leveraging the science that led to the detonation of Little Boy and Fat Man, this question could be answered. Live stars burn simple elements through thermonuclear reactions and thereby stay aloft and alive and bright. Every second the Sun burns many millions of tons of hydrogen fuel, a relentless H-bomb. All of this hot energy keeps the star puffed out and highly pressurized so that total gravitational collapse is resisted. And this goes on for a very long time. After a few billion years, when nuclear fusion is no longer energetically favorable, essentially when the star runs out of fuel in the form of light elements, the furnace cools and the outward pressure that kept the giant atmosphere aloft is no longer up to the task. The star begins to collapse under its own weight. And then what? Wheeler believed the question of the end state of gravitational collapse to be the single most important outstanding physical question of his time.

Wheeler's interest in stellar collapse in turn inspired his interest in relativity. To understand the collapse of dead stars and their final death states required an understanding not just of nuclear physics but also of gravitation, and gravitation had become synonymous with the general theory of relativity, a mathematical description of curved spacetime. Gravity would crush a dying star, but the nuclear forces would resist the compression. Which force would win?

In serendipity with the Nazi advance into Poland in 1939, J. Robert Oppenheimer and his student Hartland Snyder published a seminal paper, based on idealized conditions, in which they argued that a big, dense, dead star will collapse unrestrained, ultimately disappearing from sight. Overshadowed by the imperative of sheer survival during World War II, their work did not take hold immediately and their own expertise would be invested elsewhere. When John Wheeler turned his attention to the issue in the late 1950s, he criticized Oppenheimer's work, causing some offense. Wheeler pointed to the simplifying assumptions, which he suggested were unrealistic and led to unreliable conclusions. Collapse would not ensue unhindered to an untenable end point, Wheeler conjectured. But then he and his Princeton team, armed with their postwar insight into nuclear fission and fusion as well as some new computers, answered their own criticisms, thereby completing the catalog of the stellar graveyard.

Summarizing the decades of contributions, there are three stellar death states: Stars like our Sun will die as white dwarfs, a cool sphere of degenerate matter comparable in size to the

Earth, the pressure of densely packed electrons enough to resist total collapse. Heavier dead stars will stably end as neutron stars, an even denser sphere of degenerate nuclear matter around 20 to 30 kilometers across, the pressure of densely packed neutrons enough to resist total collapse. But the heaviest stars have no more recourse to nuclear pressures. Unhindered collapse is inevitable.

In 1963, Wheeler bounded on the stage of an assembly to lecture on relentless gravitational collapse, conceding in favor of Oppenheimer and Snyder's claims of nearly a quarter century earlier. Oppenheimer was conspicuously absent from the audience. Maybe still bruised by Wheeler's criticisms, maybe uninterested in reconciliation, or uninterested in celebrating Wheeler's contributions, he sat instead on a bench outside the auditorium, talking with friends. By then Oppie, destroyer of worlds, had interests elsewhere, not invested in his most fanciful and ultimately his most significant contribution to theoretical physics. In 1967, shortly after Oppenheimer's death, Wheeler was searching for a term to describe the ultimate dead star during a lecture, tired of repeating "completely collapsed gravitational object," and someone from the audience shouted, "How about black hole?"

(To quote Rai, "That's leaving out a lot of history, but let's leave it go.")

The collapsing star pushes past the resistance of crushed electrons, past the resistance of the neutrons. When the stellar material is compressed enough, the curves in spacetime around the collapsing mass become so sharp that even light can be caught in orbit. As collapse continues, light cannot escape the surface, as though the spacetime spills behind the crushed material faster than light can race outward. A horizon defining the region of no return, the event horizon, is inscribed in the very geometry of spacetime. The event horizon casts a lightless shadow, and a black hole has formed. The black hole is not a star anymore. It's not really even a thing. The pulverized matter that cast the shadow of the event horizon continues to fall and is gone. The black hole is nothing but its shadow.

Wheeler initiated Kip Thorne into this spectacular era of black holes and quantum mechanics. Kip was among the first generation of physicists bred on relativity. He had the good fortune to mature in a time of significant unsolved astrophysical problems that awaited relativity for their unlocking, and he had the brilliance to make the most of his good fortune.

A prize student and a faithful collaborator, Kip was also young, a child during the war, and a reformed pacifist. When I initially describe him simply as a pacifist, Kip corrects me. "Far from it," he says. "Having lived through the horrors of World War II and its aftermath and having learned of Stalin's purges, I was very far from being a pacifist." Still, his political attitudes were not aligned with those of his mentor. Kip perceived paranoia and ignorance as driving factors in the Cold War arms race. But Wheeler's controversial involvement in the escalation of the thermonuclear weapons program was an undeniable part of the intellectual context he offered. The hydrogen fusion bomb could be limitlessly powerful, a weapon of genocide. The word that came to mind when Kip thought of the superbomb was "obscene." His own interests were pure in intent: pure astrophysics, pure knowledge, knowledge that belonged to no one or to everyone, citizens of the same Earth. The superbomb is morally obscene, but the underlying nuclear physics has no intrinsic moral character. Out of a benign interest in nuclear physics, Kip asked technical questions that his friends with security clearance were unwilling to answer. His mind was on the nuclear processes driving stellar evolution, not bombs, but as Wheeler had realized, the physics was the same.

Kip put aside their political differences to admire and love his mentor for all of the reasons Wheeler was commonly loved and admired. Kip was drawn to Wheeler for his brilliance and intellectual generosity, not his politics. The magical element Wheeler commanded comes through in this quote from his (assisted) autobiography: "Now, in my eighties, I am still searching. Yet I know the pursuit of science is more than the pursuit of understanding. It is driven by the creative urge, the urge to construct a vision, a map, a picture of the world that gives the world a little more beauty and coherence than it had before."

Abstract mathematical oddities became real conquerable astrophysical terrain for Kip and his generation. Black holes were dead and dark and by some irony could churn up space and time around them to beam the brightest beacons across the universe, although any evidence for them was still disputed effectively enough in the 1960s and 1970s. Kip could delve into the theoretical details of pulsating black holes, accretion of cannibalized stars, emanation of gravitational waves. The real also encouraged thought experiments of advanced civilizations limited only by physical law and not by technology as he theorized about wormholes and time travel. Mathematical proofs would leak out of the journals and into the culture, a level of science fantasy validated by calculation. His contributions to relativistic astrophysics are foundational. It was the golden age, as Kip himself characterized the era. By 1970, at the age of thirty, Kip was a full professor at Caltech, famous, and widely respected for his detailed, thoughtful, and original theoretical achievements.

His mentor's generation was called to such vital purpose. Lives were lost and saved. A world war resolved. The scientific stakes though more abstract also had an auspicious varnish. Kip may have felt a calling, an obligation to a cause more worthy of his devotion than his own professional elevation. Kip could become an advocate, a converter, a champion—we'll stay on the atheists' side of the term "evangelist"—of a new way to commune with the universe. He could help bring a natural resource down to Earth to share with his community and inspire a movement that would in sum exceed any individual contribution, even his own. While astronomers voraciously collected the sky's light into telescopes, Kip saw an opportunity to contemplate the universe not through images from light waves, but through sounds from gravitational waves. To reference an over-referenced Pynchon title, Kip saw an opportunity to contemplate the universe through gravity's music.

I would describe Kip as careful but not cautious. His calculations are executed deliberately, unhurriedly, sometimes outright slowly. Meticulousness does not however translate to hesitance. His work is also colored with confident speculation, audacious risk, and daring. Of all the prospects Kip surveyed, he may have surmised that gravitational waves were the most thrilling, but he must also have surmised that they would be the most contentious and fugitive. Gravitational waves are difficult to understand and plagued by ambiguities. A change of perspective and the relativism of space and time might shuffle them away. Were they even real? Or just an artifact of a bad mapping of space and time?

Einstein himself was not certain gravitational waves were real. In 1916 he thought not. That same year he thought so. In 1936 he thought so again, although in between those dates there was vacillation. In a lecture he gave in the midst of this work, Einstein said, "If you ask me whether there are gravitational waves or not, I must answer that I do not know. But it is a highly interesting problem."

By the 1970s not every strain of skepticism had been cured. Still, a solid theoretical picture had emerged from the years of exchanges. Maybe not everyone was convinced that gravitational waves were real, but Kip was. He went so far as to say that by the time he started his PhD with John Wheeler in 1962, it was obvious to him that gravitational waves must exist, although the debates would continue in dwindling subsets for another twenty years. In 1972, in an annual review with his doctoral student, the accomplished Bill Press, Kip laid out a vision for the field that would guide the next few decades of Kip's career, a reflection of his late-night walk and realization that Caltech should get interested.

Conceptually, gravitational waves are required out of respect for the speed limit. As one black hole orbits another, the curves in the shape of spacetime must drag around with them, but the shape of spacetime cannot acclimate instantaneously, since that would require propagation of information—about the motion of the holes—faster than the speed of light. As the black holes move, the curves shift and adjust, and those changes wave outward incrementally and at the speed of light, carrying energy away from the violent astrophysical motions.

There are astronomical rewards promised. A "new window on the universe," as Kip has said many times, will be brought to us by these "new cosmic messengers." But details about the astrophysical events and the energy they could deliver to gravitational waves were scant. Gravity is the weakest of known forces. The gravitational attraction between two electrons is less than a trillionth of a trillionth of a trillionth of the electromagnetic interaction. The gravitational pull of the entire Earth is easily resisted by mere human muscle—we can jump. Only the most aggressive dynamism of the densest conceivable concentrations of mass and energy could send gravitational waves loud enough to ring the most sensitive instruments.

The strenuousness of the venture was countered by the prosperity of the times. The golden age of relativity encouraged a daydream of a cosmos plentiful in the unforeseen. Maybe the sonic universe would be as bountiful as the viewable universe. Galileo pointed his telescope in our astronomical yard, at the Sun and the planets. He saw mountain ranges on the Moon and concluded that the celestial body was not a godly Platonic sphere. He saw moons orbit Jupiter and rings around Saturn and deposed us finally from the center of the world. In the following centuries a proliferation of astronomical inhabitants came into focus beyond our own solar system and beyond our own Milky Way. Maybe the ifos would be rewarded with such abundance. Record the sounds from space and possibly all manner of unpredicted dark phenomena will chirp back at us, which brings us to the motivations on the day when Kip met Rai.

In 1975, Rai and Kip both made their way to Washington, D.C., for a NASA committee meeting. Kip was there to collect information in preparation for his imminent proposal for a research program in experimental gravity at Caltech. Rai recounts, "I picked Kip up at the Washington airport. I had never met him before. I thought, 'What the . . . ?' He had long stringy hair. He had ties. He had wristbands. A complete nut, okay. It was not anything I had ever seen before. He looked funny as hell to me. I probably looked funny as hell to him.

"Later we found out we were at Princeton at the same time. And I fell in love with Kip. He's a delightful man, even though he looked cuckoo, absolutely cuckoo."

Rai describes that consequential meeting: "And so what happened is we spent the whole night, literally the whole night—Kip was busy at that time thinking, 'What should Caltech do in an experimental way in gravitation?'"

Kip remembers several times staying up all night talking to Rai. "There were a bunch of nights. Beginning in the seventies. Then the eighties. And also the nineties." Kip laughs at the thought and probably some of the specifics. "But which of those times that we stayed up all night, I'm not sure. My memory is lousy."

"Because you stay up all night," I suggest.

Our conversations jar Kip's memory. He goes so far as to consult documents he has archived (being very precise and careful) and clarifies the night. By that date, he already anticipated that gravitational-wave experimentation would be a principal component of his proposal to Caltech, but the conversation with Rai may have shifted gravitational waves to the centerpiece of the program he envisioned.

Rai remembers, "We made a huge map on a piece of paper of all the different areas in gravity. Where was there a future? Or what was the future, or the thing to do? And I wasn't trying to sell him on it, but Kip came to it himself. He decided that the thing they ought to do at Caltech was, out of all that stuff, interferometric gravitational-wave detection. It looked like the most promising thing. So then there was a big discussion. And Kip was dancing around this problem of 'Well, I can't do it by myself. Whom should I get?'"

Rai continues: "Kip already had in his mind what he wanted to do. He wanted to hire Vladimir Braginsky, who is a very good person by the way. A Russian who was very close to Kip. Kip spent time in Moscow, I don't know if you know that."

Kip corrects this impression a bit, reminding me of due

process, formal search committees, involvement of provosts and academic presidents and chairs and faculty. In any case, a respectable list of potential leaders for an experimental program would have included the name Vladimir Braginsky.

There are organisms that survive despite or even because of extreme conditions, outrageous pressures and temperatures, things that impossibly metabolize just elemental hydrogen from vents on the sea floor. The Soviet scientists of that era were not such extremophiles, since the suffix implies a love for their circumstances. But they did thrive under impossible pressures and barren conditions, metabolizing the intellectual equivalent of the most basic of elements. The beleaguered but legendary centers for astrophysics in the USSR drew admirers from the West, Kip included. Kip was not particularly intimidated by the KGB's interest in his trips to Moscow. If the scrutiny weighed on Braginsky, he didn't show it, going along with the required measures with apparent patience, the vexation diminished entirely by the value of their collaboration and the pleasure of their friendship. On the few occasions that they traveled outside of the inner-ring highway around Moscow, Braginsky was required to send Kip's itinerary to the authorities so that soldiers could verify their route at scheduled checkpoints along the way. Braginsky admitted privately to Kip that each time Kip visited, Vladimir had to be debriefed by the KGB. As much as Kip was under watch on his trips to the USSR, Braginsky was under watch on his trips to the United States. Sometimes the Soviet scientists would have to travel in groups with a KGB plant among them. Kip relays, incredulous, "The KGB guy was the one who didn't know anything."

They were watched by officials on both sides. Kip is pretty sure his phones were tapped by American authorities in the late 1960s and early 1970s. On one occasion, Mr. Bevins from the

Los Angeles FBI office knocked on Kip's door for a fourth or fifth time in search of more detailed information about Vladimir Braginsky. Kip, tiring of the absurd surveillance, opened his office door: "He's right here. Ask him yourself." At which point Kip politely introduced the mute agent to his charge. After a solid pause from his shocked guests, Mr. Bevins raised the cuff of his pant leg to provide evidence—"See, I'm made of flesh and blood just like you are"—as though he only just realized the significance of the equivalence himself.

Braginsky had already convinced Kip that gravitational-wave detection was going to succeed, and Kip wanted to be involved with that prospect as more than just an advisor to the Russians. Rai explains, "Well, there were problems. Kip knew at that time it was very difficult for Braginsky to leave Russia. The Cold War was still going. I don't know how Braginsky was able to travel. But he was able to travel, which made me think he had KGB connections. But he never traveled with his wife or kids. They kept them hostage. Look, I'm making that up. But it's very plausible."

Kip assures me Braginsky had no KGB connections, although he was a member of the Communist Party, "though often not in very good standing." The travel permissions extended to Braginsky were attributable to Soviet pride. In an era of posturing and rhetoric, the Soviet authorities benefited from the respect Braginsky earned in the West. They would allow him to travel, and along the way he could display the superiority of Soviet science. Kip suggests, "Nevertheless, presumably just to make sure he knew who was boss, from to time he was denied an exit visa, and on at least one occasion his exit visa was taken from him at the airport as he was about to board the plane."

Braginsky would have been the most natural scientific collaborator for Kip to bring to Caltech, and Braginsky considered the possibility, tried to imagine a world in which he could move freely to liberal, sunny, carefree California. The brutal repercussions for those he left behind were too terrible to imagine. Braginsky stayed in Russia but even from that distance his technical influence made its way out of the country. Braginsky's group continues to impact the advanced detectors.

Rai remarks that some months after their meeting in 1975, "Kip asked me if I was interested. And I told him, 'Let me warn you right off that I have a horrible record. I don't publish and you won't get me through any committee.'

"Let me tell you a cute story. Kip persisted, and he did ask me to apply. So I sent him what I had, and he sent me back a note after I sent him my CV, and he said, 'There must be pages missing, aren't there?' Well, that was that. I figured, Don't even pursue it."

"That was actually not a big deal in our discussions about Rai at Caltech," Kip protests. "I have no doubt at all that a professorial appointment for Rai would have sailed right through the faculty and administration." (When the docket of candidates was compiled in December 1977, Rainer Weiss was second on the short list.)

But back in 1975, the night they met in D.C. before their NASA committee, Rai suggested another name that would come up again in the Caltech search process. Rai recounts, "A person that I had not known, that I had never met but I had begun to realize was a very clever person, was Ron Drever. So I suggested Ron."

4

Culture Shock

Frugality was a theme inscribed early in Ron Drever's thinking. Ronald William Prest Drever was born in a modest Scottish village, his family poor even by the economic standards of the times, although his father, George Douglas Drever, rose out of his own childhood circumstances in an industrial town outside of Glasgow to become a doctor (apparently not a wealthy one). His mother, Mary (Molly) Frances Matthews, came from a remote part of England in Northumberland near the Scottish border. She spent her childhood in a "big old rambling farmhouse," living on an inheritance ample enough that no one had to work. They could just live, though not that well, Ron admits. Parsimony was obligatory for most of his life and not entirely unpleasant.

The first house Ron's parents bought together, Southcroft, was situated on the main road in Bishopton, Renfrewshire, Scotland. The village of Bishopton had a population of about seven hundred, I learn from Ron's younger brother, John, known as Ian, who believes the house cost £200, their parents' entire wealth, drawn from their mother's dowry. At Southcroft, Ron's mother gardened avidly but made no use of her skill with horses or milking cows. Their father hung a plate out front advertising "Doctor Drever." The home became the hectic center of a country doctor's life, able to accommodate the father's medical practice with a consulting room and a dispensary, since country doctors doubled as pharmacists. Occasionally patients found Ron or his brother, or both, having a bath in the house's single bathroom. The family did not have a car. His mother did not drive. She cycled always, despite Scottish weather, as did their father, carrying out the expected home visits on rough, inhospitable roads.

There were plenty of patients, but there wasn't plenty of money. The community suffered plentifully from endemic unemployment, a failed local economy, and the anxiety of the times. Although the suffering may have encouraged poor health and visits to the village's general practitioner, there wasn't the money to pay the good doctor. Their father rarely charged his patients, whose requests for an appointment were delivered in person or occasionally by letter. Eventually some calls came by phone. Mrs. Woodrow answered the telephone exchange near the post office beside the railway station, reporting on the whereabouts of the villagers and connecting the ailing to Dr. Drever's phone at Bishopton 57. The elderly doctor in the village—"Old Foozler" the Drever boys called him—eased his practice over to Dr. Drever, intentionally or organically, but in any case gradually. Dr. Drever came to assume all official roles for Bishopton: local Treasury medical officer, police surgeon, insurance examiner, factory doctor, doctor for the post office.

Ron was born on October 26, 1931, during a difficult home labor assisted by a local midwife from the nearest town, Paisley, and another doctor also from a nearby town, both fetched

urgently as circumstances deteriorated. His father adminis-
tered anesthesia with chloroform delivered via rag and bottle. Ron was dragged into this world with application of the
dreaded and now medically obsolete forceps. Ron's brother,
Ian, wonders if Ron's general condition as a difficult child was
assisted by the forceps as well. (The instrument, medical and
symbolic, remained in the maternity bag their father continued
to use for his own practice.) Ron was "fussy," even obsessive,
and required order and cleanliness. His brother, Ian, uses an
old Scottish word for accuracy: Ron was "pernicketie." But he
was also loved and adored by his family. Ron required attention
and was given what he required, along with care and affection.

Ron's mother blamed a nanny, Willah, for Ron's pernicketie
disposition. His brother does not. He thought Willah great fun
and says, "The problem was not Mother or Father, or Willah,
or anyone, but something in Ronald's gift of personality." Only
after leaving home, becoming a doctor himself, did Ian exam-
ine the vortex around his older brother. "I was unaware of the
anxieties Ronald caused until I was at school myself, and realized that the world, our world, revolved around him."

Ron's brother recalls the incidental, intimate details that
color a childhood: "Luckily, a boyhood and longtime friend . . .
gave Father a loan of a bullnose Morris car. This was a great
success; only one problem was that it had no doors to get in,
which meant swinging legs over the side followed by a climb.
Mother found this awkward, being regularly smartly dressed,
if visiting. . . . I have no memory of the bullnose Morris, but
heard so many stories about it, a wheel all by itself running
past them after rounding a bend beyond Dumbarton, and they
all laughed at some poor other road user losing a wheel, when
their own car lurched horribly and collapsed—it was their own
wheel; trips planned to wherever the Sun was shining, regard-

less of distance or difficulty, picnic parties to the Trossachs, wild narrow roads, areas down the Clyde Coast to visit with wonderful aunties, uncles, friends."

Ron's uncle, John Richan Drever, known as Rec ("an unmarried man," Ron says), was an artist, but given the demand for fine art in an economically depressed Scotland, he went into shipbuilding as a backup. (Ian tells me the Richan family married into the Drever family when both were farmers on the Orkney Islands. When the Vikings invaded they conferred derogatory names on the local inhabitants. He says Richan meant "dregs" and Drever meant "rubbish.") Rec ("a joyous, massive support," says Ian) lived with the family for a time while writing for newspapers and enrolled in correspondence courses for commercial art. Ron learned all of his practical skills from his uncle, investigated motors and engines, found odd tools, and honed an artist's care for fine carving.

Ron would repair clocks and radios for his father's patients, many of them offering metal fragments or pieces of wood for him to play with instead of toys. He struggled with his letters at school but fared brilliantly in science. While at the Glasgow Academy, his class built a television from "surplus junk." Ron headed the group of students that equipped the set with sound. Later, he made his own television in the Southcroft garage, from which family and friends watched the queen's coronation in 1953, the blue screen only a few inches across. It may have been the only television in the village. His brother remembers, "Ron made a radio-controlled gadget . . . chased, sniffed at by a puzzled cat." Ian still keeps the tiny electric motors that Ron built in a small tin intended for windup gramophone needles.

During World War II, their father, affected by his own service during the previous world war, determined to keep his family close, although the modest village was not spared the fighting. A large munitions factory built on a nearby marsh attracted German bombs that would land in the mud without detonating. Later the British army recovered the devices and brought them elsewhere for controlled explosion. Occasionally the Drever boys could collect fragments of shells and casings pitched around the garden from engagements overhead.

"I was in charge of Ron. I was unconsciously trained to keep an eye on him. I learned always to be with him," his brother, younger by three years, explains without resentment. "We have always been close. You couldn't be angry with Ronald, he just wouldn't understand." Later, the brothers traveled together by bus to the University of Glasgow, as they had to the Glasgow Academy. "Our parents' concern for him never ceased." After Ron earned a degree, they encouraged him to decline a research position at Cambridge University, "anxious how he would manage daily life." He did decline. "Anyway," his brother reassures me, "Glasgow was the best place in the world according to Ronald."

Ron liked to make things from nothing, from random bits of rubber tubing, sealing wax, scraps of previous experiments, anything lying around the university lab or even the house, or, in one notable case, some materials from his mother's garden. He admired thrift, and the lean budgets of his early endeavors only enhanced his pride when the things worked with precision and efficiency.

While at the University of Glasgow he struck on a peculiar idea that he could perform nuclear experiments using the Earth's magnetic field, a kind of natural nuclear magnetic resonance probe. "It was a very strange, weird one—very unusual," Ron explains. He strung things up in his mother's well-tended garden using an accumulation of car batteries from the family garage and some equipment borrowed from the students' laboratory at the university. He stayed up for twenty-four hours in the back of his family home in the quiet Scottish countryside with an old camera (old at the time of the experiment even) and "an ancient scope," taking measurements every half hour on lithium nuclei in a jar of solution. Essentially, he was testing Mach's principle, which, very roughly, asserts that matter out in the universe affects things as fundamental as inertial mass down here. He was interested in a specific variant of Mach's principle that suggested that the distribution of matter in our own galaxy, which is roughly a spiral in a plane with a dense center, would affect the inertial mass of the lithium nuclei in the jar of solution. As the Earth spins over twenty-four hours, his mother's garden would rotate relative to the galactic center, the densest region in the Milky Way. Ron set out to test if the nuclear properties of lithium changed—an indication that the inertial mass changed—with motion and orientation relative to the plane of the Milky Way. Apparently there does not exist such an effect, a perfectly good answer. The setup was scrappy and rudimentary and that was just fine with him.

He learned that another group published results of a similar experiment involving laboratory-grade magnets, but Ron thought, "I can do it. The stuff I've got costs nothing." He wasn't discouraged by the competitive advantage that a sophisticated laboratory must have had. Quite the opposite. He didn't need an expensive magnet. He had the Earth, and the magnetic field the Earth provided was free. In the end, he published an assessment of his experiment, "and it was a bit more sensitive than this other chap's with the more fancy equipment, and mine cost essentially nothing—just a few car batteries and some wire. It was fun."

The Hughes-Drever experiment, named after Ron and "the other chap" (Vernon Hughes from Yale), is now considered a precision test of the principle of equivalence, which suggests that matter will fall freely and weightlessly in a gravitational field. (Admittedly this isn't the most conventional phrasing for the equivalence principle, but it captures the relevant aspect for this context.)

On the basis of this peculiar but ingenious experiment, Ron was offered a fellowship to Harvard, where he and his advisor, R. V. Pound, worked out some clever experiments that I won't go into here. He says of his transition to Harvard, "I had a kind of provincial existence: I didn't travel; I had no spare cash. I didn't go abroad for holidays, because of work. It was pretty near the first time I was out of the country. So that was a great year. I found the whole place astonishing and so different from what I had expected."

After his productive fellowship at Harvard, Ron returned to Glasgow with more experience, some grant support, and a modest group of researchers. He would tinker around the lab and rearrange the tools, often alone and somewhat bored, almost excavating the empty air for ideas in pursuit of something exciting. His observations synced with the dark phases of the Moon, he'd travel into the heavier dusk to dawn of the countryside, receptive to pale astronomical flashes, the Moon redirecting the sunlight away from the Earth, the less luminous messages straining at visibility. He joined collaborations with like-minded explorers, and his interests broadened. All the observations yielded negative results—they saw not much—but were no less interesting to him for that. He was surveying the skies before he drilled deeper. His interest shifted away from light to the prospect of noise. Gravitational waves bounced along the communal consciousness, fomenting people's interest. The more Ron listened to colleagues around the United Kingdom—Hawking and Sciama and Jelley and Aitken—the more he suspected the waves were real and detectible. Gradually ideas were exchanged, simple implements were devised, radical but accessible experiments were modeled, and one step was taken after another until some direction and enthusiasm grew.

Ron was proud to cut rubber matting right from the floor of the lab and stack the offcuts with leftover lead bricks to build crude and simple components that worked well. He could construct something precise and powerful from nothing with nothing but his bare hands and glass cutters and windowpane, bits of paper, rubber bands, stray screws, and this he describes with such pleasure and amazement. He had nothing and he could make something admirable of it.

By the time Kip approached Ron with an offer from Caltech in 1978, Ron had already designed his own ifo in Scotland with ambition and austerity equally influential. He wanted to build the instrument on a "shoestring" but as big as they reasonably could. The University of Glasgow scrapped a synchrotron, a kind of particle accelerator, and Ron managed to reconfigure the abandoned space for an ifo more than double the size of any other in the world, but still only a quarter as long as the one Caltech promised.

As Ron tells it, Kip urged him to consider as enticement the deeper financial support for basic research offered in the United States in general and at the world-class Caltech in specific. Caltech was exciting, but Ron defended Glasgow's scientific reputation. Underrated, he implied. In Glasgow he could do as he liked with almost no red tape and less money, a constraint that may have appealed to him. And the Glaswegian project looked competitive. But nothing could quite stand up to Caltech.

He struggled with the decision and sought the advice of

his allies in Glasgow, lifelong supporters. ("I have the highest feelings about them," he says.) They encouraged him to follow the opportunity. Still, unable to make the decision, he struck on a five-year trial period during which he would split his time between Glasgow and Caltech. He says, without tension, "I didn't realize at that time that things were very different in the U.S.A. than they were in Britain, and in ways that were nonobvious at the time to me. And that people thought differently and acted differently. . . . I didn't realize that, that they were so different."

I've only ever heard Ron Drever's Scottish cadence in recording. Generally he sounds easier than I expected, mildly upbeat, with kind words for the history makers. If I sense any difficulty in his temperament, it is only in the subtle resistance to assent to even simple statements. When Caltech hired Drever in 1979, he was known for clever ideas and an obvious talent in experimentation. He was inventive, dedicated, and equal parts impossible, as soon became apparent. Rai Weiss has horse sense, according to Kip, and in retrospect Kip admitted to me that he wishes he had given the value of that character trait more consideration in the early days.

Ron cultivated the aura of a scientific Mozart—Rai's analogy—a childlike spirit attached to a wondrous mind that just seemed to emanate astonishing compositions. Everyone around him was forced to play Salieri, unfairly catalogued as a plodding technician in the shadow of Mozart's genius. Talented scientists felt miscast by Ron in the lesser role. The laboratory was his private Exploratorium. Before he came to an agreement with Caltech, Ron insisted that he would go to Caltech only if he was in charge of the project himself, if he could run it any way he liked; "I thought that was understood." He roamed that Exploratorium unconventionally but creatively, in a gale of his own thoughts conjured up in pictures more than equations. This ability to intuit without recourse to ordinary logic no doubt lent that extra magical aura to his reputation as a genius, so that others felt a bit prosaic for their habit of actually calculating from a supposition all the way to a conclusion. But there was an incapacity associated with Ron's disposition too.

By all accounts, Ron would release a deluge of ideas on his team each new day. Ideas were abundant. But decisions were scarce. The next day, the joy of unhindered exploration would begin anew, and he would release another deluge of ideas on his flustered team. Progress was as random as the walk of a shred of lint through hot air. While Ron was in his native Scotland, back at Caltech nuts and bolts were tightened, only to be furiously loosened on his return. Meanwhile, in his absence, his crew in Scotland would move on with their own experiments knowing full well that soon Ron would reappear and reverse all of that forward-directed motion.

To provide continuity while Ron was back in Scotland during alternate months, Stanley Whitcomb was hired as junior faculty in 1980 to oversee design and construction of the Caltech lab in a space that hugged the machine shop, closing off windows and putting an end to sunlight in a strip of rooms along the shop. Stan also provided expertise and more, an adept intuition as the delicate character of ifos as a species would emerge. (Rai: "Stan is solid as a rock. He's a fantastic guy. Smart as hell.") Ron would parachute in with his flurry of ideas, but on a day-to-day level, Stan ran the team that built the system, pulled the vacuum, installed the laser and crude mirrors for an operational Caltech prototype by 1983. Initially, the R&D facility was designed to demonstrate the effectiveness of Drever's design, to test the stability of the laser, to probe the sensitivity of the setup. Ron wondered if they might even expect detection capability, although Stan claims agnosticism at the time. Ron's uninhibited optimism may have been an inflated version of a general optimism in the early 1980s that sources would be plentiful and the sky would be loud. If they are plentiful, they have not been loud. (Kip assures me that reports of "general optimism," at least among the theorists, are highly exaggerated. He cites his article in a 1980 volume of *Reviews of Modern Physics* in which, yes, he did say that the laws of physics did not prohibit loud sources, but he also said that the astrophysical understanding of the time pushed the gravitational waves much quieter, not too far off the current LIGO target—a characteristic signal about a billionth of a trillionth of the length of an ifo arm.)

Ron would design the instruments on the eleven-hour flight between Glasgow and California. Covering notebooks with detailed drawings, he'd work out practical tricks implemented by Stan Whitcomb at Caltech or Stan's Scottish counterpart, Jim Hough, at Glasgow. While Ron was away, Stan would push things forward at Caltech and Jim would push things forward in Scotland. Ron admits this must have been a bit hard on his two groups.

The instruments were noisier than expected and led to several ingenious solutions, which ranged from seismic isolation to stabilization of the laser, cleaning the light, recycling and building up the power. Ideally one instrument would have been more advanced than the other so that they could better exploit the R&D opportunity, fixing a previous generation's weaknesses, staggering the efforts between the countries. Instead the instruments were nearly identical, with Glasgow implementing phases just marginally ahead of Caltech.

After the five-year interim, Caltech as well as the NSF wanted some security in the form of a commitment from Ron. Caltech pressed Drever to choose: Stay or return to Glasgow. Ron thought the arrangement was quite satisfactory, maybe unaware of the discontent on the ground, happy for the quiet up in the air on the long flights, grateful for the chance to arrive with a new design sketch for the technical draftsmen at Caltech.

He had to choose between the promise of Caltech and the comfort of Glasgow. The atmosphere was good in Scotland, a culture he better understood, that he perceived as more open and collaborative. He felt the scientists, even the students, at Caltech were discouraged from collaboration, were unwilling to work together, were more competitive.

"I afterwards realized that this was a general thing; that people didn't want to work together—they all wanted to be independent. That was a major difference, which took me a long time to understand, and that was one of the problems later on, I think." When Shirley Cohen interviews Ron Drever for an oral history in 1997, she challenges him: "But look, Ron. On another level, you came, and you didn't want to share being in charge with anybody." To which Ron replies, "No, I didn't." And Shirley Cohen presses, "So in some sense you could understand people wanting to be independent." And Ron says gently, "Oh, yes. Perhaps that's true." Although I didn't get the impression he thought it was at all true. The problem he perceived was a defect of American culture, and as evidence he did point to his successful and friendly alliances back home. (According to those in Glasgow, the collaborations were profoundly less ideal than Ron remembers. People describe his conduct as fraught, competitive, suppressive even.)

Eventually though, Ron did realize the project needed to grow. He imagined something gradual with intermediary machines, but bigger nonetheless, and Caltech seemed more likely to succeed in that upgrade than Glasgow did. The decision was made. In 1983, he accepted a full-time position at Caltech, with regret that he may have abandoned his team in Britain.

Kip had finally secured the campaign at Caltech. The stakes were high. Unknown fundamental physics. Uncharted cosmology and astrophysics. The magnitude of the avowed objective would not have been lost on Ron. The pursuit of gravitational waves, a somewhat obscure subject not central to mainstream astrophysics, had become a major new initiative—today the biggest single undertaking ever at Caltech.

Ron pushed his design forward with the Caltech prototype. There were tensions between him and Rai, but Rai was back at MIT with his own modest prototype that Ron dismissed— "unjustly," Rai would say—as burdened by inferior technology, less financial means, and other drag forces Ron didn't care to identify. The German group was still a consideration. According to Rai, they deserve the accolade for the finest of all the prototypes. Ron's former group in Glasgow would carry on without him, potentially breaking out as competitors too. Although the Glaswegian team lost an unusual talent with Ron Drever's departure, they also gained unfettered access to their own lab, finally able to get on with things. But Ron had Caltech, he had Kip, and he had the largest ifo in the world, which on compilation would leave behind any potential competitors. Nobody in the world could possibly keep pace with his ideas; he felt fairly sure of this already. Add to that natural advantage the leverage of his new position and his new laboratory, and the future was promising. The machine was his. The ideas would be his. The decisions would be his. By rights, the future of gravitational-wave astronomy was his too.

Ron failed to incorporate the influence of human psychology on his plans. That there would be hurdles to the future was inevitable. The job definition was to flexibly and ingeniously catapult over technological obstacles. That they would be hampered by the past was also inevitable.

A controversial history would subtly shape the project, possibly more than any of the key players had anticipated. Before there was Rai, Kip, or Ron, there was a lone contentious pioneer, Joe Weber, who left the theorists to their debates, heads down, hazardously close to their chalkboards and books, and decided instead to go out and have a look around. If gravitational waves were real, he would be the first to find them. He struck out alone, a brazen explorer, and came back with tales of a remarkable view he didn't claim to understand yet. Many would kit up and follow him, adrenalized by the promise, although acclaim quickly turned to ignominy. Scientific disputes broke out and dissipated the brief warmth of attention.

Joe Weber errantly persisted for thirty more years, relentlessly, doggedly—the treasure too valuable, the defeat too personally catastrophic—caught up in his determination and unwavering purpose, caught up in aspirations for the improbable if not downright impossible, caught up in ambition, but not for money, just for knowledge and acclaim and respect. Caught up in a luckless expedition.

5

Joe Weber

In 1969 Joe Weber announced that he had achieved an experimental feat widely believed to be impossible: He had detected evidence for gravitational waves. Imagine his pride, the pride to be the first, the gratification of discovery, the raw shameless pleasure of accomplishment. Practically single-handedly, through sheer determination, he conceives of the possibility. He fills multiple notebooks, hundreds of pages deep, with calculations and designs and ideas, and then he makes the experimental apparatus real. He builds an ingenious machine, a resonant bar, a Weber bar, which vibrates in sympathy with a gravitational wave. A solid aluminum cylinder about 2 meters long, 1 meter in diameter, and in the range of 3,000 pounds, as guitar strings go, isn't easy to pluck. But it has one natural frequency at which a strong gravitational wave would ring the bar like a tuning fork.

Yonah was born in New Jersey in 1919 to a family of brothers and Lithuanian Jewish immigrant parents. "Yonah" became "the Yankee" became "Joe." A schoolteacher unable to interpret his mother's accent misheard "Joseph," and his mother nodded that that was close enough. Joseph Weber could have been Yonah Geber, but the family accepted passports filled out cheaply and quickly in the name of Weber.

Joe Weber dropped out of Cooper Union to save his parents money, joined the Naval Academy, and became an officer, a radar expert, a navigator, and eventually a commander. He was on the aircraft carrier USS *Lexington* when it sank during his World War II naval service and eventually commanded a submarine chaser. He tells Kip Thorne in an interview Kip recorded in 1982, "I was the one given the job of finding the right beach to land Brigadier General Theodore Roosevelt, Junior, and eighteen hundred rangers in July of 1943. After the war I was head of the electronic countermeasures section . . . so I knew electronic countermeasures of the entire navy." Joe's accent has that raw Americana I associate with men of his generation. His family called him "Yankee" after a childhood accident—he was hit by a bus at the age of five—required speech rehabilitation that broke his Yiddish and scarred him with a broad American accent.

After his service, he was hired as a professor at the University of Maryland with "one of the highest imaginable salaries. The princely sum was sixty-five hundred dollars a year." He was twenty-nine. Oddly, he had no PhD, although it was a condition of his employment that he procure one. To this end, he approached the famous physicist George Gamow for a PhD project. Professor Gamow asked, "What can you do?" Joe explained, "I'm a microwave engineer with considerable experience. Can you suggest a PhD problem?" Gamow said, "No," according to Joe. Just "no." The irony that he doesn't have to explain to Kip but that I ought to explain here is that Gamow with Ralph Alpher and Robert Herman predicted the existence of the light left over from the big bang, currently lingering in the microwave bandwidth—the cosmic microwave background radiation, the glow from the origin of the universe. Had Gamow said yes, Joe and he might have won the Nobel Prize for its observational detection. Instead, in 1965, by accident, almost begrudgingly, Penzias and Wilson, scientists at Bell Labs and subsequent Nobel Prize winners, detected the primordial background light. It's been said that Penzias and Wilson essentially observed the most important thing anybody has ever seen.

Now replay: A young aggressive expert microwave engineer goes to the acclaimed Gamow, who predicted microwave radiation left over from the big bang, the single best evidence for the origin of the universe, and asks, "Do you have a project for a microwave engineer?" and inexplicably Gamow says, "No."

Joe's scientific life is defined by these significant near misses. After Gamow's mystifying dismissal, Joe went on to study atomic physics and thought of the concept for the maser (microwave amplification by stimulated emission of radiation or, in more modern usage, molecular amplification by stimulated emission of radiation) and by 1951 gave the first public presentation, which had all of the essential ideas. So Joe is credited, though not as widely as some argue he should be, with an independent discovery of the predecessor to the laser (light amplification by stimulated emission of radiation). Had his luck been different, he might have shared a Nobel for that discovery too. And the patents. And the money. Joe had nearly made the summit before. He was Shackleton many times, almost the first: almost the first to see the big bang, almost the first to patent the laser, almost the first to detect gravitational waves. Famous for nearly getting there. After his disappointment with masers, he mentions casually to Kip in the interview, and with no irony,

"One of the reasons that I wanted to, ah, get into relativity research was that it didn't seem to be a field with, ah, any particular controversy."

In black-and-white photos he bends over the cylinder, hair both dark and grey, brushed back and high, shirt white and short sleeved, glasses black and square. He affixes quartz crystals to the middle of the bar, which when squeezed by the bar's resonant vibrations produce an electrical voltage that sends currents through the electronics wired off the bar's midsection to record the harmonics of the plucked string. The contraption is modest and demands little. Weber had a bar at the University of Maryland, in an ordinary-looking laboratory, occupying a sensible footprint in a little room managed easily by one person. He built and installed other bars about a mile from campus in a structure that could accurately be compared to a garage. Weber then stationed a bar at Argonne National Laboratory near Chicago, far from the Maryland bars, to impart confidence to coincident events, to rule out neighborhood mayhem, car crashes, and storms. He has thought this through. He has been ingenious and tenacious and daring. The bars are cheap. And they work. Daily they ring in response to multiple signals from our galaxy. The universe rewards him with a noisy sky. He doesn't presume to identify the sources. He's agnostic about the sources. He leaves that to the theorists. He has discovered a new frontier for experimentalists to explore and theorists to explain. He has made one of the experimental discoveries of the century. It takes him and his small team a decade, a respectably extensive investment, yet less than the hundred years projected by skeptics for successful completion of a viable experiment.

In 1969 at a typically uneventful conference on general relativity, the kind of conference where the very existence of gravitational waves was still under debate, Weber makes his announcement. He's detected "Evidence for Gravitational Waves," as his paper was titled, maybe from colliding stars, or neutron stars, or pulsars, somewhere over there, around the center of the galaxy. There is shock, then celebration. Applause even. He is heralded. He is on the cover of magazines. He is famous.

Kip remembers the announcement, and while he was surprised Joe presented results so soon, he thought to take them seriously. Excited by Weber's reports, physicists tried to understand the sources capable of ringing his bars so consistently and so energetically. Theorists were also inspired to contemplate all manner of novel sources, not necessarily to explain his experimental data so much as to explore the full terrain of possibilities the universe might have to offer. Roger Penrose considered crashing gravitational waves. Stephen Hawking threw black holes at each other. Enthusiasm paled as calculations came in. Weber estimated that the energy output from our galaxy would have to correspond to the destruction of thousands of Suns yearly to be consistent with his data. His agnosticism on the sources is appropriate for an experimentalist, who should remain impartial and unbiased. But to a theorist, that sounded like an implausible bounty of energy. Martin Rees, now Sir Martin Rees, showed with his collaborators Dennis Sciama and George Field that the amount of energy Weber claimed to observe was simply too much energy for our galaxy to produce without fragmenting. Still, there were uncertainties in the calculations, and Weber persisted unflaggingly, accepting the ambiguities left open to him.

Joe spent some time with John Wheeler in Princeton, where he first met Kip along with the influential theoretical physicist Freeman Dyson. Weber and Dyson discussed the possibility that exploding stars—supernovae—might ring spacetime, and

Weber may have chosen the resonant frequency of his bar to align with this possibility. Joe mocked theorists for their superior attitude, but despite himself he valued Dyson's encouragement, recounting, "Dyson wrote, said he'd been thinking about it. First when he heard this project was being carried on he thought I was insane but he had thought the matter through and had done the first gravitational-collapse calculations and he had sent me those and those are reproduced in this book *Interstellar Communication*."

In *Interstellar Communication,* a sober if unconventional look at the merits of communication with prospective extraterrestrials, Dyson published the article "Gravitational Machines," in which he pursued the promising candidates for gravitational-wave sources: compact dead stars. Although we're able to see compact dead stars by now, in 1963 their existence was uncertain. Dyson imagined that an advanced civilization could arrange two orbiting compact dead stars to slingshot spacecraft to near light's speed. He also recognized that such a pair occurring naturally would emit a powerful burst of gravitational radiation that Weber's equipment might detect. Dyson's idea survives, not as a form of extraterrestrial communication but as one of the most promising sources for a direct detection of gravitational waves.

Weber reads Kip some motivating lines from the publication: "Freeman J. Dyson, 'Gravitational Machines.' . . . 'Loss of energy by gravitational waves will bring the two stars closer together with increasing speed until in the last seconds of their lives they plunge together and release a gravitational flash of . . . unimaginable intensity . . . should be detectible by Weber's existing equipment. . . . Seems worthwhile to maintain a watch for events of this kind using Weber's equipment.' "

Even the great Oppenheimer encouraged Weber. Joe picked

Oppenheimer up from the airport during a visit in the mid-1960s and found him to be very enthusiastic about the pursuit of gravitational waves. Joe relays Oppenheimer's remark: " 'The work you're doing,' he said, 'is just about the most exciting work going on anywhere around here.' I was astonished and of course it gave me a tremendous morale boost. He was not one to give a compliment lightly." So that's the early history, he tells Kip. Evidence registered, labeled, filed.

As fast as scientific momentum can change, it changed. Soon there were resonant Weber bar detectors under development at IBM, Stanford, and Bell Labs, and in Scotland, Japan, Germany, Italy, the Soviet Union, California, Louisiana, and Rochester, New York. All over the place. Literally all over the place. In 1972, NASA even put one of Joe's instruments, the lunar surface gravimeter, on the Moon. There were new designs and refinements and analysis techniques. And no one, no one except Joe, detected a gravitational wave. It was dead quiet out there.

Ron Drever, still in Glasgow at the time, and his collaborators, as well as other groups around the United Kingdom— from Harwell, Cambridge, Oxford, Glasgow—devised their own bars with ingenious alterations and enhancements beyond the simple tuning fork. Ron began his investment in the bar technologies in the early '70s, motivated by the belief that Weber could be right.

Stephen Hawking and Gary Gibbons of Cambridge discussed equipping a lab from parts literally culled from a junkyard, although the assemblage never materialized. Ron investigated a tank in a junkyard on their behalf, its original purpose to decompress divers. He concluded it was indeed cheap but useless.

Ron requested to visit Weber at his Maryland laboratory

sometime in the 1970s, but Weber, angry and suspicious, replied flatly that Drever was not welcome. Ron arrived anyway and found the contempt undiluted. Weber's greeting: "You can't just walk in off the street and do gravitational wave experiments." Ron quite agreed, but somehow Weber couldn't discern his upbeat tones. Unfazed by Weber's reception, Ron returned to the United Kingdom to build and expand his own bars in Glasgow. Although he had reason for doubt, he permitted an optimism that rendered his mind open to the prospects. To his dismay, the bars offered only noise, and quickly he and his collaborators acknowledged that the doubt had ripened to a conclusion: Weber must be wrong.

Braginsky had been the first to build a bar and announce a negative result: no waves. He ran the experiment for only a few weeks and quickly decommissioned the apparatus to try harder with better bars or consider an entirely different experimental approach. Ron Drever's bar experiments were next and more extensive. He also spent a year or two trying "all kinds of crazy ideas," and there was talk of a big British project at Rutherford's laboratory in Cambridge. The German group also refuted the claims of detection with a very reputable bar experiment. Bars were edging to the periphery.

Joe Weber's claims in 1969 to have detected gravitational waves, the claims that catapulted his fame, that made him possibly the most famous living scientist of his generation, were swiftly and vehemently refuted. The subsequent decades offered near total withdrawal of support, both from scientific funding agencies and his peers. He was almost fired from the University of Maryland. Weber summed up his circumstances with a self-effacing remark about his second wife, Virginia Trimble, a young astronomer twenty-three years his junior. The sociologist Harry Collins recounted, "[Weber] told me with a smile that when he married her he was famous and she was not, and now their roles were reversed."

Weber never relented even as the evidence against him accumulated and the community turned away. Although occasionally his claims of a direct detection of gravitational waves are reexamined, the evidence weighs heavily toward the negative. Weber never measured a gravitational wave but instead recorded a fault in the equipment, or made an error in the analysis or interpretation, or worst of all he unconsciously biased the data.

Richard Garwin, an experimentalist from IBM, initially motivated by Weber's claims, or motivated possibly by suspicion, briskly fashioned and calibrated his own detector at the narrow frequency of Weber's original bar, and when the skies were as quiet for him as they had been for the other experimenters around the world, he was displeased. Convinced from previous exchanges that Weber would not be swayed by reason or raw data, Garwin opted to confront him publicly, ambush style. At a relativity conference at MIT in 1974—these conferences were getting progressively more lively and driven by controversy, kudos to Joe—Garwin stood at the front of the hall and denounced Weber and his accomplishment. Weber and Garwin almost came to blows, the two tense in threatening postures in front of an audience of otherwise peaceful relativists and separated by the upheld cane of a polio-sabotaged astrophysicist, Phil Morrison. They retreated, Weber resolute, Garwin contemptuous.

Joe was goaded, pulled deeper into his convictions. Garwin's experiment was not as good as Weber's, some might say. It was smaller, thrown together briskly and operational only for a month. And in any scenario, two different experiments are never identical and require painstaking efforts to compare.

As a scientist, Weber had a right and more, an obligation, to indicate the failings of the comparison. He couldn't concede his own failure on the basis of incorrect logic and poor data.

Circumstances did not get better for Joe for the next twenty-five years. His detractors caught him in egregious errors. He made claims that were indisputably false. Joe noticed that when the center of our galaxy was overhead, once every twenty-four hours, he recorded clusters of events. He deduced the signal could be coming from the dense galactic nucleus, where lots of gravitational activity might plausibly generate a significant output of gravitational waves. The astronomer Tony Tyson was in the front row alongside Johnny Wheeler and Freeman Dyson during a colloquium in Princeton in which Weber showed a plot with a big peak in the data once every twenty-four hours when the galaxy was overhead, indicating powerful bursts of gravitational waves coming from the dense galactic center. "We all leapt to our feet," Tyson recalls, "and said, 'Wait a second, Joe, gravitational waves should pass right through the Earth.'" Problematically for Weber's conclusion, since gravitational waves pass right through the Earth, his bars should accumulate events once every twelve hours, when the galaxy was overhead *and* when the galaxy was underfoot. When the error in reasoning was pointed out to Weber, he reanalyzed the data and came back within a couple of weeks with clusters of events every twelve hours. This flexibility in his data analysis deepened distrust.

Tony Tyson had built his own bars at Bell Labs and thought he better cut his losses. He ran his bars for over a year and "didn't see a damn thing." But there was an excitement still surrounding the possibility of new physics, enough excitement that he couldn't resist the urge to do better, push the limits farther. David Douglass at the University of Rochester built a

carbon copy of Tyson's bar so they could search for widely separated coincident events. Courtesy of AT&T, Bell Labs' parent
company at the time, a direct coaxial cable ran between Tyson's
lab, Douglass's lab, and Weber's lab. They could each directly
download the others' data onto digital tapes and perform separate analyses.

In an analysis of the data from the independently built
and operated bars, Weber claimed there were glitches coincident with those registered by his detector in Maryland. Coincidence across widely separated and independently operated
instruments would support his claim that the signals were truly
astrophysical and not local earthly interference. Meanwhile
Douglass and Tyson found nothing above the noise.

Tyson conjectures that Joe was pulling false signals out of
the noise of his own data coincident with entirely fake pulses
Tyson intentionally injected into his data just for the purposes
of calibration. "I thought we informed Joe of the calibration injections. Maybe we didn't," Tyson explains, mystified.
If Joe claimed coincidence with these false signals, he could
detect coincidence anywhere. Even more damning, the groups
recorded the data using different time standards. Tyson and
Douglass used Greenwich Mean Time, while Joe used Eastern
Daylight Savings Time. When Joe recorded an event at 02:00
and claimed simultaneous events were recorded by Tyson and
Douglass's equipment at 02:00, the events were actually offset
by four hours. There were no simultaneous events. Under the
best of circumstances, it would be impossible to recover from
that kind of mistake. Eventually, Joe removed himself from the
task of data analysis to quell any accusation of his personal
bias, but it was too late. People became unkind. Intentionally
he was tricked and deceived into false hopes that seemed to generate false claims, the ruse revealed in very public, very humili-

ating forums. He became a fraud to expose. Tyson says of Joe, "He was a great electrical engineer but a lousy statistician."

By the late 1980s the professor emeritus used his wallet to maintain his laboratory, an unembellished concrete box between a Maryland wood and a golf course. He is reported to show the actual wallet to make the point. The sign in front he did not maintain with particular care, and its pride faded, the announcement of the monument, "Gravitational Wave Observatory," blanched by weather.

6

Prototypes

There's a building on the Caltech campus that looks like a trailer from a certain angle and is hard to find even with an iPhone and a map annotated with GPS coordinates and an arrow pointing to the precise longitude and latitude. I make it as far as an industrial lot and then wander past an unimpressive alley that leads to the only entrance to the "40 meter," the colloquial name for the Caltech ifo prototype and the building that houses it, a sort of attachment to Central Engineering Services. The electronic map-plus-arrow confusingly marks a nonplace as *the* place, a location neither on the street nor properly rooted to campus.

I leave my target unnoticed a hundred yards behind while I take some grief from Jamie over the phone, which glows with my current coordinates, for not finding the 40 meter despite his conscientiously documented map. "I'll come get you," he says, trying to sound more provoked than he is. "Go back the way you came." So I do, retracing my steps into the industrial garbage area.

Jamie Rollins, Rai Weiss's former graduate student, worked the 40 meter prototype in recent years before shifting to different posts within the collaboration. Backtracking two or three steps, I point to the trailer door off a loading lot, the gesture intended as a question. I catch up to him and make a show of looking confused. He feigns bafflement. "I gave you a map."

A single door marks the threshold of this makeshift structure, not actually a building but a temporary enclosure slapped together thirty years ago, a shelter to test and develop the experiment. It is true central operations are primarily maintained in a trailer. But the lab has to fit 40 meter tubes in two orthogonal directions, so by the laws of reality cannot possibly be contained within any facility with the dimensions of a truck. I haven't walked around the perimeter, but clearly some other structure is melded onto the trailer front. The decades of labor hidden by this unassuming single-story building will culminate in an accomplishment that I am going to try to impress upon you deserves new adjectives, new descriptors. I'm crossing the modest threshold of a trailer into the R&D phase of an experiment that will measure waves in the shape of space less than a billionth of a trillionth of the length of the machine.

Proportions alternate between infinitesimal and astronomical. The signals are infinitesimal. The sources are astronomical. The sensitivities are infinitesimal. The rewards are astronomical. The human ambition to understand the universe is merely epic, and astronomical trumps epic.

The building and the 40 meter prototype belong to no one person. The entire team can be replaced and has been piece by piece assembled and reassembled as scientists come and go, and the machine still hums, indifferent to its attendants. Countless students have trained in the narrow confines, and the directorship of the 40 meter lab has rotated over the years. Each element of its anatomy has been designed, scrutinized, built, tested, improved, documented, dissected, and then integrated into the rest of the body. The best ideas are implemented and vivisected and debated and restructured and eventually installed in and commissioned for the two full-scale LIGO observatories at the remote locations (not at Caltech or MIT), where the elements have to be rescaled and retuned. None of the physical labor can be executed at a particularly brisk pace, and the experimentalists learn a certain patience, move with a special unflustered clarity, acquire a long-term view. They maneuver around the model not quite space-station slowly, but sure-footed and unhurried. The laboratory itself is sometimes empty, with more warm bodies in the control room monitoring the internal workings of the ifo on a typical day.

At the Caltech 40 meter Jamie tosses me a pair of protective glasses and dons a pair himself that look oddly similar to his actual corrective lenses. We put on some paper booties to keep street dirt on our shoes and off the lab floor. The booties come in two sizes, effectively big and small. I am directed to the small pile. Costumed in booties and glasses, we walk past a narrow hallway of a room occupied by a couple of graduate students and a couple of postdocs in the proverbial hunch over computers, past a narrow control room with black-and-white monitors trained on the optics, and then through double doors to enter the actual domain of the instrument. There are signs over the doors and on the doors and on the side of the doors. I intuitively recognize them as hazard signs, but I don't take in the content. I write Jamie some weeks later to remind me of the exact phrasing, and I get this in reply:

There are a lot
a lot of words

DANGER
RESTRICTED ACCESS
LASER
Etc.

The warning signs create the perfect tone so that a visitor won't just stumble into the lab fecklessly. A proper entrance requires care and deference, as though the workroom were part china shop, part factory. There's sticky tape on the floor to keep the lab that much cleaner, the tape glomming onto the bottom of booties, extracting the dust and dirt accumulated on the short trip from the bootie basket to the swing doors of the lab.

An experimental physics lab is probably unlike any other room you've been in before. The lighting is harsh, of course, aggressively bright and beyond the reach of aesthetic concerns. There are sounds of machines, a harmonic hum, sometimes just from fans on computer equipment as opposed to any motorized parts. There's never any bespoke sound absorbers, so the machines have a sonic clarity that seems intentional, cranked up for some postindustrial experimental orchestra.

There are two stainless steel tubes, each 40 meters long and about a half meter in diameter. The steel tubes, better known as arms, are arranged with respect to each other in an L shape. Wires are dangling in certain places, and the space to walk alongside the tubes is pretty narrow. I'm a bit anxious in the lab and my glasses keep falling off.

"How good are the filters on these glasses?" I ask, thinking they might be unnecessary.

"Very. If even one photon scatters out of the beam into the room and then into your eye, you're totally fucked. [That's a wild exaggeration, I'm assured.] Wear the goggles. We brought the vacuum up in the optics chamber to atmosphere, and I've had my whole head in the instrument all week. It's very stressful."

I spend the rest of the tour with the protective glasses pinned on my face by one hand.

By the early 1980s, Drever brought all he had to the Caltech 40 meter prototype. The very thing about the 40 meter that is no longer true for anyone was once true for Drever. The 40 meter was his. "I thought that was understood," Drever had said.

Drever never considered ifos proprietary, that ifos belonged to Rai or to anyone else. In fairness, neither did Rai. Rai is quick to say that a careful sifting of the history shows he wasn't the first to consider the utility of ifos as gravitational-wave detectors. In the 1970s, "there was in the United States one other guy who was working on this—a student of Weber's whom I knew about. His name was Bob Forward. . . . And he was chasing an idea. You see, this idea . . . is not just mine. Other people were doing it."

Drever says in the 1997 interviews, "There was a chap called Robert Forward." Forward at the time was working at Hughes in Malibu and managed to convince the company to let him build an ifo of his own.

Rai remembers, "Forward had gotten the idea from a guy I had talked to. And he claims that it came from me, through a guy named Phil Chapman to him. I don't think that's quite accurate. I think it really came from Weber. Weber also had thought of using interferometry as a way of detecting gravitational waves.

"Well, it turned out that later on we learned this amazing thing. Kip did some research and found out that two Russians at Moscow State University [Gertsenshtein and Pustovoit] had published this idea already in the Soviet *Journal of Experimental and Theoretical Physics,* back before I had even thought about it. And I didn't know about that. I don't know their names, but they are the two names that appear in some of our writings now. They had a crude idea, which was an idea similar to what Bob Forward and Weber had—namely, using light as a way of measuring the distance carefully. So, you know, interferometric detectors grew up in a lot of places. The significant thing I did was to actually see how practical the idea was by doing this noise analysis, which I felt was crucial. And that's not being modest or anything."

Rai explains his frame of mind in those days. "It wasn't a completed experiment, it was an idea, and you didn't publish a thing like that. But there was a piece of me that said that it ought to be put someplace, so I put it into the quarterly progress report . . . a big, long report. And then that was it; we didn't publish it anymore. And in there was the basis of, in fact, the whole damn thing."

Kip independently mentions the importance of Rai's analysis of the noise, which in this context just means anything that rattles the machine, from traffic to earthquakes to quantum fluctuations in the laser light. Kip calls the report a "tour de force." The basic idea is there in a very clear form in the original Soviet paper by Gertsenshtein and Pustovoit, but no noise estimates, no assessment of the feasibility. "The idea becomes a reality with Rai," Kip emphasizes. "Rai identified all the dominant noise sources that initial LIGO would ultimately have to face and devised ways to deal with them and did a noise analysis for an instrument that incorporated those ways. It was far more of a tour de force than Rai gives himself credit for. In retrospect, it really was amazing."

Rai tells me, "I didn't publish because it was an idea. And I still hold to that. I can't have it both ways now and I'm wanting it both ways. That's the trouble. This is important as a philosophical point. I don't know how you feel about this, but it's very different to have an idea than to do it. They're different things. I get pissed off endlessly when some guy or girl has an idea and publishes it somewhere and they own it and they haven't moved a goddamn finger to make it happen. And they haven't done the hard work to slug it through. The person who deserves the credit and who publishes it should be the one who made it happen."

In the '70s, Ron Drever had to find Rai's internal MIT quarterly progress report on microfilm and enlarge an almost illegible photocopy of the document. The British group considered lasers among their crazy ideas, but Drever was uncertain where the idea had come from, if it came to them through the ether, the Germans, or Rai's original report. The Glasgow group operated with very little money, and lasers seemed prohibitively expensive. The laser itself could cost nearly £10,000, and that seemed exorbitant, enough to steer them off that direction for a couple more years. Drever too had seen Rai's NSF proposal and given the project a very strong endorsement. Billings in Germany soon got started on ifos, and although they had thrown around some ideas involving lasers before, the realism of Rai's proposal knocked these groups into action.

Drever had been thinking through the technology for ifos for two years and was in the early phases of the Glasgow operation when he met Forward in the '70s. Unsure of the true origin of the idea of a free-mass ifo as a gravitational-wave detector, Drever indicates more vaguely that ifos were in the air. A key idea, however, that Ron attributes to Rai is the realization that bouncing the light in an arm multiple times would significantly boost sensitivity. Drever came up with a different version of this idea, the Fabry-Pérot cavity, that, naturally, would be much cheaper. Ron felt himself in a friendly rivalry with the German group, which seemed to have more money, more support, more everything, which he admittedly envied. The Fabry-Pérot cavity that Drever struck upon was going to give him a competitive edge. He presented the idea at a conference in Jena in East Germany, still behind the iron curtain of the Cold War. He says, "It was quite an excitement going through the curtain, and so on, on the bus and seeing how miserable the place looked compared to West Germany."

From these early beginnings, interferometers grow in intricacy as each group contributes systems over decades. I've only seen simple line drawings of an interferometer before. A line drawing is to a real interferometer what hangman is to human biology. A real interferometer is a material manifestation of decades of research and breakthroughs and tinkering and quite a lot of basic labor. I could never make this climb. I'm one of the bespectacled crew at the foot of the mountain shouting suggestions, theories, cheering the climbers on with their practical shoes and ice picks and tools. Here find a theorist's admiration for the physicality of the experiment.

At the Caltech 40 meter I'm looking at a ridiculously detailed schematic of the instrument, almost a poster, affixed to the wall of the lab, alongside one of the arms. The professionally drafted plan illustrates the complex architecture of the prototype. Many more crossed lines than I expected to be necessary trace the trajectory of laser light through the apparatus, which in the cartoon representation I've seen before is rendered (inaccurately) as a simple round trip. The schematic is entirely unreadable to me. I consider preserving the technical drawing in a photograph for later perusal, but I think better of it. To join the LIGO Scientific Collaboration (LSC), you have to sign a memorandum of understanding according to which you legally commit to the expectations of the collaboration. I haven't signed an MOU so I'm not sure what's allowed.

"If I'm not officially an LSC member, why am I allowed to see all of this?"

Jamie is amused. He nudges me: "What, are you going to run home and build one?"

If you were to run home and build one, this is what you would need to do. Find a seismically quiet site. Then build two tunnels in an L shape, the longer the better. As a gravitational wave moves past, distances are stretched and shrunk by some minuscule fraction. The magnitude of the gravitational sounds detectible depends on the length of the tunnels, better known as arms. The change in the arm length due to a typical gravitational wave may be less than a billionth of a trillionth of the length of the arm. Make the arms too short and you're going to be less sensitive to gravitational waves and you will not notice the modulations of space.

At the corner of the L, position a powerful high-energy laser. Shine the laser light into a beam splitter, which as the name implies splits the beam, sending some of the light down one of the arms and some of the light down the other. Crucially, pump all of the air and contaminants and everything particulate out of the tunnels so that the light travels unhindered through empty space. That is one of the major challenges, to keep the arms empty. No air should scatter, absorb, or otherwise interfere with the laser light. The light will make its way down an arm of the full-scale instruments in about a hundred thousandth of a second.

At the end of the arms, hang very good mirrors by the thinnest wires. A mirror so suspended can move very nearly freely in the transverse dimension. If space is oscillating, the mirror will be free to bob on that wave along the tube direction.

With those fine mirrors, reflect the light back down the arms from whence they came to recombine back at the apex, the light from one arm interfering with the light from the other at the beam splitter. The light recombines perfectly at the bright port and cancels perfectly at the dark port, if the beams have traveled exactly the same distance. If instead one of the arms was getting shorter while the other was getting longer, the split beams will not have traveled exactly the same distance and when they recombine there will be an interference pattern that documents the modest difference in the distance traveled, one moving a ten-thousandth of a proton more or less than the other, amounting to differences in their travel times of one thousandth of a trillionth of a trillionth of a second (an octillionth of a second). Now you have a working interferometer.

Build. Repeat. Because you need two. At least two. Build the second one far from the first. The second serves not only to confirm the detection as real and not a false alarm, but also to ascertain the location of the sound. The utility of two detectors on the Earth is like the utility of two ears on the head.

So there's the cartoon. Build an L. Establish a vacuum. Shine a laser. Hang some mirrors. Recombine the light. Detect the interference. Record the sounds. Easy.

Easy.

You'd almost have to believe it was easy. Rai Weiss started with the simple idea: Let mirrors float freely in space and roll with the waves. Build an interferometer around the freely floating mirrors. At one of the observatories recently Rai stormed out of the main laboratory during part of the Advanced LIGO installation, cursing murderously. A colleague in witness to the storm, unsure whether he should say something, landed on a lackadaisical "How's it going, Rai?"

"The machine is too fucking complicated!" Rai shouted without stopping or looking for the source of the query. "It's too fucking complicated."

7

The Troika

The moral tone of academic censure might sound familiar, even if the specific codes defended are peculiar to a given group. Among scientists, it is practically criminal to be wrong. Verifiability is substratal to any scientific endeavor. There is nothing to be gained and everything to be lost with false claims of an experiment that rings for you if the apparatus can only return dead silence for the rest of the world. Weber must have believed his statistical interpretation of his data. It's not my view that Weber intentionally misrepresented his results, it's not even the prevailing view, but it's also not an entirely eradicated view.

When Rai Weiss and Kip Thorne and Ron Drever surveyed the aftermath of the ruination of bar detectors, there would have been every incentive to back away from the whole mess, to keep a safe distance from any splatter in defense of their pristine scientific reputations. Instead, each of them, and independently, considered the wreckage and believed there to be a treasure of great value in the direction that Weber had signposted. Ron took some time to consider the geography before turning to ifos. Rai was hooked the first time he found ifos in a corner of his mind. And Kip patiently accrued the advice of the experimentalists before setting his course.

Kip quotes Einstein in his book *Black Holes and Time Warps:* "The years of searching in the dark for a truth that one feels but cannot express, the intense desire and the alternations of confidence and misgiving until one breaks through to clarity and understanding, are known only to him who has himself experienced them." Among their misgivings must have been Weber's quick rise to fame and painfully public descent. Many experimentalists abandoned the field. The broader community was definitely not generously disposed to an investment of actual dollars in new technologies for the lark. At least Weber bars had been cheap. Ifos were not going to be built out of rubber matting cut from the floor or spare batteries from a Scottish garage. Risk aversion had taken hold. To many scientists, the search for gravitational waves was dead.

But Kip and Ron and Rai were gripped by "the intense desire" to continue, to slog and struggle for "a truth that one feels but cannot express." They worked hard, "the years of searching in the dark," far more years than any of them anticipated. They strove for a glimpse of that greatness—to break through "to clarity and understanding." Black marks and all, the Weber history an unavoidable part of the context; they were snared. The competition between the groups only drove them harder. None of them could turn back. The only direction in their field of view was toward the summit.

As Ron Drever and Stan Whitcomb built the 40 meter at Caltech, with Kip's powerful presence and theoretical efforts mounting, Rai went in his own direction. Although an open communication remained between the Caltech group and the MIT group, theirs were still separate operations of very different prototypes based on some distinct technical ideas. To Ron at least, he and Rai were somewhat competitors. Ron was impressed by Rai early on, and I'm sorry that in the quote I can't effectively convey the pace and the particularly Scottish laying of stresses: "Very early on he had a lot of *stuff*, he had vacuum tanks, a laser, all the main things *made*, very very early. The strange thing was he just didn't seem to *move* from there . . . for *years*."

Rai didn't have a lot of money and he didn't have a lot of support. He says, "I remember vividly trying to tell people in the department why I wanted to look for gravitational waves and one of them was to look for black holes, and people said there aren't black holes. Forget about it.

"That played a very important role in why LIGO was not started at MIT. Some of the faculty did not actively campaign against it, they were friends of mine, but influential people spent much of their lives right up until the end believing that everything that was evidence for a black hole could be interpreted without the black hole. And that set the tone at MIT. The atmosphere had been poisoned completely. MIT was not a place that was friendly to modern gravitation."

Rai's first students to write graduate theses on gravitational waves received hostile receptions from their dissertation defense committees. Their 1.5 meter ifo prototype was never going to be sensitive enough to record the clanging sounds of true astrophysical sources. They wouldn't even hear the Sun blow up. A committee member quipped something to that effect: We could do better by looking out the window. Rai is still pissed off, still gums a visceral "bad taste." Tremendous ingenuity went into both the technology of the prototype and the foresight to develop algorithms to understand hypothetical data. One student searched for exploding stars, another for black hole collisions. True, the instrument was still too insensitive by a factor of a million to actually detect these sources, but they had laid out the plan for the future. "The boys were still deeply in this mind-set: Where is the physics result?" Rai and his students couldn't make scientific claims. They could say nothing about the actual astrophysics.

Rai never believed that detection-capable ifos could be built small. He knew too much about the physical limitations that manifested as noise. Student after student pushed the floor of the noise down, giving a chance for a real signal to compete with the background racket. But still, noise was hundreds of thousands or even millions of times louder than any expected sounds of space. With each examination the projected scale of a viable detector grew with certainty. Rai knew he would never build another prototype. He wanted to go for science. He labored against his own instincts and was forced in a direction he resisted. He knew firsthand the wasted effort that went into massive projects, the headaches, the management nightmares. But the science coerced him into a big project, a huge machine— not 1.5 meters or 3 meters or even 40 meters. A machine in the range of many kilometers was the only realistic option. "I didn't like big science. But I could only do [the experiment] if it was a big project. There was no other way to continue. The science required it. I didn't ever believe you could build it small."

By late 1979, Rai had maintained his prototype for about a decade, had extracted all he could from the little model. He headed to Washington, determined to convince the National Science Foundation program director for gravitational physics, Rich Isaacson, whom many credit as singularly responsible for keeping the project from going under. The foundation had already put some modest funds into the development of the new generation of machines. Money went to Drever at Caltech and less went to Rai at MIT. He dreaded the expansion. In his little lab, they could make everything by hand. The upsurge would be expensive and time-consuming. But the experiment required substantial ground on a remote site, a more complex instrument, in every aspect an extensive undertaking. And Rai needed the collusion of the NSF.

The program director, Rich Isaacson, was "very honest and a hell of a messiah for this thing, and why was he a messiah? Because he had worked in this field himself." Isaacson was one of the first to prove, through a convincing formal calculation, that energy was lost to gravitational waves. Isaacson was deeply invested scientifically and as a program officer wanted this for the NSF, this wonderful venture, because nobody else was going to claim the territory. Gravitation as a discipline was not under the purview of the Department of Energy or the Department of Defense or even NASA. Isaacson saw an opportunity for the NSF to acquire a dramatic new vista all their own. They were promising a new brand of astronomy, gravitational-wave astronomy, a means to record aspects of the universe invisible to telescopes. Gravitational-wave detection was risky, controversial, technologically nearly impossible. Gravitational-wave detection was also a singular path to becoming much more interesting as a foundation.

Isaacson and Rai liked to walk Walden Pond, a short drive from Rai's home. The arbiter between money and science would stroll in collusion or sometimes dispute with one of his charges, possessed as though lives were in the balance, not mere knowledge. On this particular visit Rai went instead to D.C., and I'm not sure if there were habitual sceneries to wander near enough to NSF headquarters, in lieu of an exchange over desks, but I like to imagine the conversation camouflaged by a District of Columbia proxy for Walden Pond. The two men would walk in the open air like spies concerned over enemy espionage made easier indoors. Rai explained his experiences with the prototype, the inherent limitations, and the community resistance. Isaacson's enthusiasm for the scientific potential aside, the obstacles were prohibitive. The cost was uncertain but even coarse estimates encouraged the thought "staggering," on the order of the entire budget for the field of astronomy. And in the rest of that field, the blot of a scientific catastrophe (a scientist had been wrong) stained opinions. Weber's legacy guaranteed opposition from the very community Isaacson was responsible for cultivating.

Rai explains, "I used to visit Weber's lab all the time in Maryland. We were sort of unfriendly friendly. But I give Joe the credit, and I certainly tell that to his wife now, that he deserves the credit for having started the field. He was imaginative, but he was not a good experimenter. He certainly made it hard for everybody after that.

"The trouble with Weber and that whole history was really very serious to them, okay?" (Although I should interject to say that Isaacson contends the impact of the Weber history is exaggerated, compounded by repetition, and was less constraining at the actual time decisions hardened.) Rai said to Isaacson, "Well, I can't continue with it unless it does become something that does science."

Rai offered to conduct a thorough study in collaboration with industrial partners to determine feasibility and costs of a scientifically viable ifo—an industry study, not a scientific study. If the results were compelling and a full-scale instrument was achievable, then Isaacson would have the documents to support a case for a new big science initiative. If the prospect was good, then Rai would gather a consortium of scientists, all the people in the world still excited by the prospect of gravitational-wave detection. "I will collect them together, I promise you, I told him."

If the study was not compelling, they would all move on. At least Rai would. I imagine them shaking hands to seal the deal.

Rai and his MIT team invested three years in the study, called "The Blue Book." Nearly ready to submit their findings to the NSF, Rai ran into Kip and Ron at a meeting on general relativity held in Italy. "My son went with me. It was the first time I had ever taken my son anywhere. He was thirteen or fourteen. And I remember the Germans were there, the Scots were there, the Drever group, and Kip was there. And we began to discuss the idea of how they would join in at the end in this study.

"And so Benjamin, my son, went with me to meet Kip and Drever. I tell them about the plans from the Blue Book and Kip, by this time, had been working on me, saying, 'Look, why open this up to everybody? The bar people are not interested in this thing. And we're very interested.' He wanted to make it not a multi-university thing but a two-university thing, Caltech and MIT. And here's where I'm a little shaky. I don't know why I agreed, but I did. In part I agreed because—well, I had this tremendous respect for Kip. I still have that—a love and a respect for him. He suggested this, and I thought it would be—I don't know. It had to do with my personal relationship with Kip, in a way, okay?

"I didn't know Ron Drever at all; I didn't know how complicated he was until that night at that hotel. And then I realized, all of a sudden, that I was dealing with a person who was totally off the wall. I kept telling him about this plan and how we would have to do this together, and he absolutely resisted. He said, 'I didn't come to Caltech to work with you. I want to do my own thing. Why do I have to work with you?' You know, that kind of thing. That kept up all night long—I mean, for most of the evening. And my son was sitting there, and he couldn't believe his ears. And Kip was trying to calm the thing down. And Ben told me later, 'What are you trying to do? This guy doesn't want to work with you. What the hell are you trying to do?' So I said, 'Well, he can't do it himself. I can't do it by myself. This is just too big a thing. We've got to figure out a way to do it together.'

"Well, that problem never really ended. I may have a year screwed up here, but anyway now what happened was that Kip convinced me that it ought to be a Caltech-MIT joint report." (Although, for historical accuracy, Kip emphasizes, "the Blue Book was a strictly MIT document, with some individual contributions from Caltech's Stan Whitcomb.")

Rai continues, "And we made a presentation—in October of 1983, I believe—finally."

Rai says, "I was scared to death of this whole proposal, which was presented as an idea that would cost about seventy million dollars—an outrageous amount of money; that's what came out of the industrial study. . . . It had two sites, but it was bare-bones in every other way. And Ron Drever was drawn into this by Kip, kicking and screaming. He wanted no part of it; he wanted to do his own study. He wanted to do everything himself. And Kip tried to convince him, although Kip didn't know that much about how big science gets done either, that it wasn't going to be done by him alone.

"Well, no. That's where the story's all screwed up. Kip and I don't agree on this; you're going to get a different version from him. I knew it couldn't be done by one institution alone, okay? I had to convince Kip that that was the case. . . . Kip makes it sound like it was a shotgun wedding at that point. But the wedding was inevitable. I knew it. It was ridiculous to think otherwise, and the NSF said so in the end. It took them long enough.

"And, by god, a lot of them were thinking of Nobel Prizes. Okay? That's the sin in this field, if you want to know the truth. Yes. I think it's the key thing. That is, if I have to pick a thing that made Ron Drever so impossible, it would be that. That's my own take on this. I tried to accuse him of that in Washington once, but he'll never admit it. And the NSF would talk about this being one of the things that would happen. If this happened, it would be a new field, and consequently they would be responsible for a Nobel Prize in physics. Okay? And that's very important to an agency. So I think that's the background.

"All right. Where were we? Okay, we have this rickety arrangement that's barely holding together between Caltech and MIT. It's really not between Caltech and MIT, it's between Ron and myself.

"And what happened very quickly at Caltech is that when they saw that MIT was faltering—I mean, Caltech can act very fast when they want to; MIT can't.

"So what happens is that MIT does nothing. They're very glad when Caltech jumps in and effectively takes over the project. And I got mad as hell. Okay? You might as well know it. And I've been mad as hell about those people forever. Not the people at Caltech. They're the saviors of this thing. I'm really bullshit at MIT."

(Rai says later, "That was all true back then. But not now." MIT is a collection of people, and with a change of administration in the mid-1990s, the whole mind-set changed. "MIT has been very supportive of the project ever since. That's important.")

Drever's move to Caltech was not planned in the context of a massive merger with MIT. He imagined that sheer cleverness could deliver a small interferometer that was scientifically viable. Rai by contrast knew that wasn't practical. Ron just would not concede. "He never lived in this world anyway." Kip understood as Rai did. Their alliance was going to make the achievement conceivable. Kip tried to convince Ron, but Rai says, "He had this child on his hands. This brilliant child."

Ron complained about Rai. "I felt that he was trying to knuckle in to the project that we were trying to do, and he was also always very competitive. . . . There were all kinds of arguments, although we got on in a friendly enough way." He protests, "Rai Weiss had had an early idea, as I said, on this. He was doing experiments—in my view, very slowly—on a very small scale. And they weren't going very fast. While I was working at Caltech, the stuff we were doing at Caltech was way ahead, in my view, of what Rai Weiss was doing, was also going much faster, and was much better in practically every way. Rai had a very small effort. Basically, he made a small interferometer and it really didn't work very well. It wasn't the fact that it was *small* but that it wasn't designed very well, in my opinion." (Obviously Rai disagrees.)

Ron was "shocked" and deeply worried when he heard about the Blue Book study. He felt that the small-scale problems had not been resolved and that no one was ready to enlist industry in a massive, staggeringly expensive project. And certainly he didn't think Rai should be leading the charge. Rai's ambitious plans for a huge system irked Ron, who instead wanted to grow in steps with intermediate-scale machines that became progressively bigger. But Kip insisted that the jump to large-scale interferometry was vital. A sequence of intermediary machines just for the purposes of R&D, each too insensitive to make a detection, simply would not fly. Over the course of many conversations, the standard lore is that the NSF told Drever point-blank: Combine efforts or it's over. And although Ron would visit the NSF repeatedly, as Rai had, the foundation was impassive. The next increase in scale had to be capable of a detection, and a project of such a massive scale would only succeed, certainly would only be funded, as a bigger collaborative effort. The merger was inevitable.

Ron became rather annoyed, specifically with Rai. "He was against, basically, all of my ideas. He was kind of knuckling in on the thing and wanting to try and do it differently. . . . What annoyed me also was that in the meetings we had together, he would come up with all kinds of fancy plans, and so on, with dates and so on for doing things, and I felt, well, he's trying to run the show. And yet the techniques that worked were the ones that we'd developed. And I didn't like this." Ron adds, "The ideas were my ideas."

But Rai attributes his confidence in the technology to the exemplary work of the German group. Their prototype was the best in the world. They understood their apparatus entirely. They achieved the anticipated performance that Rai predicted in his obscure progress report from 1972. All told, he credits the Germans with providing the justification for his Blue Book study. By the time of the fraught conversation between Kip, Rai, and Ron at the meeting on general relativity in Italy, the MIT gravitational research group, led by Rai, had already completed the industry study with engineering firms that tested the design of certain elements and essentially priced parts. All aspects of the ifo were examined. The tubes, the buildings, the laser, the sources were thoroughly interrogated by those industry partners, and after three long years Rai synthesized the results for the specifications with his MIT colleagues Peter Saulson and Paul Linsay in the 419-page Blue Book. "A Study of a Long Base-line Gravitational Wave Antenna System" was submitted to the NSF in October 1983. The estimate was just under $100 million for two instruments in the kilometer-scale range, bare-bones.

The instruments would not be built on that budget. The estimate was still too low by hundreds of millions of dollars. But the conversation had finally begun to break into reality.

The abstract of the Blue Book reads, "The positive conclusion of this study may have been anticipated. It could have been otherwise: the basic concept could have been flawed, the technology could have been inadequate, the costs could have been beyond reasons. None of these appears to be the case."

The Blue Book did not in any way guarantee grant support, nor was the industry study itself a proposal. The Blue Book did rigorously establish the viability of the experimental goals.

Rai and Ron and Kip, over the months subsequent to the submission of the Blue Book, struggled to formulate and agree on a development plan. Compelling presentations to the NSF followed. Kip inspired them with the promise of astrophysics. Ron charmed them in his lovely brogue with dreams of a creative design, a spinner of beautiful tales. And Rai cinched the net with the concrete conclusions of the industrial study. The central point was established. They could do this; they could build a device to record the skies.

Soon after, the project finally had a name: LIGO. Rai takes responsibility, both in the sense of credit and in the sense of blame, for the name. Kip wanted to call it a "beam detector." Rai found the affect too sci-fi. He came up with something else sitting at his kitchen table working out acronyms: the Laser Interferometer Gravitational-Wave Observatory. The O in LIGO cost them later, caused them "unbelievable pain," and practically drove them under. But that grief was to come several years later, and then in front of Congress.

Rai scrapped the little 1.5 meter ifo and built a 5 meter industrial prototype to develop components for the field. The instrument operated in wing F of the Plywood Palace until the week the building was demolished. Today a bionic reconstruction lives next door to the honeycomb of offices where Rai shows me his relic Altec Lansing speaker from the Brooklyn Paramount. Scientists crawl around the now-massive frame, amputating the old and grafting in the new technologies.

No formal arrangements had been made, but Rai says, "What happens was we make the impression we are a group, the three of us—Kip, Ron, and myself. We [eventually] made a strange organization, the Troika."

Kip assures me, "There was a lot more to it than that." A more detailed account would trace the formation of the Troika to pressures that extended over several years culminating in autumn 1983. "The creation of the MIT-Caltech collaboration was very complex. And tortured," I am apprised.

Among these history-shaping initiations, alliances, and shifts in power, one pivot would be most vital: The game was on.

8

The Climb

The astronomer Jocelyn Bell Burnell says of Ron Drever, "He really enjoyed being ingenious." She came to Glasgow from Northern Ireland to study physics and arbitrarily was assigned Drever as an undergraduate advisor. He would tell his few advisees the most interesting ideas on his mind, including those that led to the Hughes-Drever experiment (although she hadn't realized he executed this one in his rural family yard), none of which helped them pass their exams. After her initial frustration that he would not assist them with their solid-state physics homework, she came to appreciate his deep insight in fundamental physics in conjunction with his remarkable gifts as an experimenter. Drever in turn was to be influenced by the imminent and vital discoveries of his former undergraduate student. He says of her, "She was also obviously better than most of them. . . . So I got to know her fairly well." He wrote references in support of her job application to the key radio astronomy facility in England in the mid-1960s, Jodrell Bank. But, he relays, "They wouldn't take her on, and the story was that it was because she was a woman. But that's not official, you see. So she was very disappointed." He adds, hoping the absurdity was obvious, "Her second best was to go to Cambridge. You see?" He considered this a very fortuitous and happy turn. He laughs. "So she went to Cambridge and discovered pulsars. You see?"

Later in her career, Jocelyn Bell Burnell moved into X-ray astronomy to work on the team that built the British-American Ariel 5 X-ray astronomy satellite. On October 10, 1974, early in the morning, Ariel launched successfully, and at noon she heard the announcement of the Nobel Prize for the discovery of pulsars. There were two aspects of the announcement that were of particular significance to her. For one, the Nobel committee had finally acknowledged astrophysics as a subfield worthy of the Nobel Prize in Physics. In the 1920s Edwin Hubble had campaigned for such a shift unsuccessfully. For another, she was not among the recipients. The prize went to Antony Hewish and Martin Ryle.

As a twenty-four-year-old graduate student at Cambridge, she and her advisor, Antony Hewish, were looking for quasars, bright radio sources that looked as small as stars. At the time that she was stringing radio antennae in the field, quasars were still called quasi-stellar radio objects and the sources were a mystery. The radio antennae worked well at finding quasars, poorly at resolving their sizes, and brilliantly at changing the course of astrophysics. Among the quasars detected were many glitches and peculiarities recorded on the reams of chart paper, quantified by the length of paper in feet. She examined hundreds (thousands?) of feet of paper meticulously. Most of the anomalies were attributable to human-made sources or some form of detector interference. But one funny signal persisted. She became convinced that the source was astronomical in origin. She said the realization that she had seen something truly important came gradually. As is often reported, the regularity of the signal earned the sources the internal nickname of LGM, for "little green men." It turns out that there are even more precise clocks than those manufactured by the civilizations of little intelligent green men. And those would be pulsars.

Pulsars are highly magnetized, rapidly spinning neutron stars. The strengths of the giant astronomical magnets are millions, trillions, or, in the extreme, thousands of trillions of times the strength of the Earth's magnetic field. Neutron stars, always less than twice the mass of the Sun and less than 30 kilometers across, can spin rapidly, roughly once a second or even many hundreds of times a second. Particles accelerated near the speed of light along the magnetic field radiate to create a lighthouse beam that sweeps around as the nearly perfect sphere of condensed nuclear matter spins. Famously, a teaspoon of a neutron star would have the mass of a terrestrial mountain. The gravitational pull is so strong on its surface that a person would essentially liquefy and merge with the nuclear packing of the star. As a result of the strong gravitational effects, a neutron star is intolerant of imperfections. The gravitational pull overcomes any escalating mountainous formations. The imperfections on a typical neutron star's surface are so small that a 10 centimeter irregularity qualifies as a mountain, although this depends on unknown details of the neutron star crust. They spin very regularly, creating an eerily timed periodic signal in the data stream. As the beam whips around and sweeps over the Earth, the effect is of the ticking of an extremely precise clock, in some cases more precise than the most accurate atomic clocks. Of course when Jocelyn Bell Burnell discovered the first pulsar in 1967, all she could deduce for certain was that there was a very regular series of pulses, a little over a second apart, and that they were coming from the sky.

When the second one appeared in the data, "that was the sweet moment," she says. That's when the oddity began to take on the features of a discovery. "Once I'd seen one scruffy signal, I was open to seeing more." She found the first four pulsars ever discovered by human beings.

A year later a pulsar was discovered in the center of the Crab Nebula, a luminous remnant ejected during a supernova explosion. The Crab Nebula was seen from Earth and noted in historical records as an astronomical event in 1054 AD. The implication: Neutron stars are the collapsed core that remains after a dying star explodes. We now extrapolate that there are hundreds of millions of neutron stars in our galaxy, and hundreds of thousands of these are pulsars.

Hewish need not defend his credibility as a Nobel laureate. As the advisor he set his student to the task—even if the task was to look for quasars. Harder to comprehend is the omission of Jocelyn Bell Burnell from the list of recipients. I ask her if she thought her former advisor should have done something more, and she says with no resentment, "If you get a prize, it's not your job to explain why you got the prize." She also adds that the slight has worked out for her quite well. She continues to get seemingly every other prize, medal, honor, and accolade ever invented. Fair compensation she seems to imply. Dame (Susan) Jocelyn Bell Burnell: dame commander of the Most Excellent Order of the British Empire, fellow of the Royal Society, president of the Royal Society of Edinburgh, fellow of the Royal Astronomical Society, many distinguished medals, dozens of honorary doctorates etc., etc., etc.

Pulsars took sides in the theoretical debate. The lighthouse beacon of neutron stars had been spotted out there in our own Milky Way, a few hundred light-years away. The half century of speculation about the end state of gravitational collapse, the question Wheeler had prioritized beyond all others, had brought astronomers to this juncture. Pulsars were the first bit of evidence that neutron stars were real. If neutron stars formed as the death state of stellar collapse, then black holes might too. Einstein dismissed black holes, not yet bearing the name, as a mathematical solution of interest but of restricted applicability. Matter would resist such catastrophic collapse. The architects of nuclear weapons came to understand otherwise. A massive-enough remnant after the violent cycles in a star's demise would succumb to unhindered collapse, blow past the neutron star state, and continue to fall, leaving a black hole in its wake. But there is nothing like plain observation to finally resolve a theoretical standoff. Jocelyn Bell Burnell found evidence of a neutron star. Added to the sheer intrinsic fascination of that discovery was the promise of even more, the promise of black holes. (An illustrious colleague is reported to have intercepted her at a meeting to declare, "Miss Bell, you have made the greatest astronomical discovery of the twentieth century.")

Although pulsars made black holes seem more plausible, observers would patiently accrue data over decades before most everyone would agree. There is a real astrophysical black hole in the direction of the constellation Cygnus, an arbitrary configuration of stars, as are all constellations—some of the stars that share the borderline of the same constellation are thousands of light-years deeper than the others. The stars align deceptively when falsely projected onto the surface of the sky. They align with enough fortuitous organization that Ptolemy connected the dots into a pattern of barely essential details to evoke a "swan," the meaning of the word *cygnus.*

The black hole gets a name of its own, or a derivative of the name of the constellation. We call it Cygnus X-1. And this catchy label indicates the direction of the black hole and the nature of the discovery, because astronomical names need an informative purpose. The black hole is in a binary, which is to say the dead star is not alone. It has a companion, a live big blue star. The binary emits copiously in X-rays, extremely high-energy light. X-rays are energetic enough to penetrate your soft tissue but not energetic enough to penetrate your bone. You could take a picture of your skeletal structure with the light coming off Cygnus X-1.

Discovered in 1964, the black hole in the constellation Cygnus was arguably the first black hole ever discovered. But the controversy about the viability of gravitational collapse to the total catastrophe that is a black hole was not resolved until the 1970s, with a minority holding out until the 1990s. A big blue star—a supergiant—orbits very close to the black hole, which is about fifteen times the mass of our own Sun. Atmosphere off the blue supergiant blows down onto the black hole, the wind falling into an orbit and creating a thin disk of material circling the hole and gradually dribbling across the event horizon. The black hole is slowly consuming its companion, and in the process the infalling matter gets heated to millions of degrees, and when things get hot, they get bright. The whole region surrounding the black hole is powerfully glowing in X-rays emitted by the infalling material.

In reality the pair is about 6,000 light-years from the solar system, uncorrelated in physical location with the actual placement of the other stars in the constellation, as claimed, united only in direction. The black hole and blue supergiant execute a full orbit about once every five days. The amazement never ceases.

Some excessively cautious astronomers might still call the compact object in Cygnus X-1 "the putative black hole," "the alleged black hole," "the conjectured black hole." We cannot see the black hole, just the effect the curved spacetime has on matter, and we extrapolate that the object at the center of the viscous spill of hot matter siphoned off the blue supergiant is so small (about 88 kilometers across) for something so heavy (again, at least fifteen times the mass of the Sun) that it must be a black hole. Those inordinately prudent observers are few, granted, but the point is pressed. We have never *seen* a black hole.

The quasi-stellar radio objects Hewish and Bell Burnell set out to investigate (later called quasars when their extragalactic origin became apparent) looked bright and small, like stars, but were scattered outside of the plane of the galaxy, which was a hint that quasars don't actually live here in the Milky Way. They are on the order of a billion light-years away or more, which means they are old, the light traveling billions of years to get here, and rare, which means the universe doesn't make them as much as she used to.

Quasars are the energetic output of the entire core of an ancient galaxy shining bright enough for us to see at considerable distances. A supermassive black hole (the putative, conjectured, alleged) millions or billions of times the mass of the Sun drags the galactic analogue of driftwood—entire stars, gas and debris, denizens of the astronomical galactic nucleus, the ephemera of the conglomeration's formation, all of it—into a hot mess tumbling down into oblivion. The black hole can spin up the miasma into a luminous jet propelled millions of light-years out, a cosmological signal we saw from Earth for the first time in the 1960s and didn't know what the hell could possibly do that.

Quasars are one kind of active galactic nucleus, all of which are powered by supermassive black holes. With the mass of a billion Suns concentrated in an area the size of our solar system, the active galactic nucleus is a heavy anchor, accumulating a dense and populous center. There could be tens of thousands of smaller black holes and other dead stars and some live stars orbiting the nucleus. The supermassive black hole may have its own origin in the seeds of dead stars, in stellar-mass black holes that collided and merged and grew to be the elephantine core of the galaxy.

Very nearly everything we've learned about this specific universe, the details of its actual landscape and inhabitants, its history and morphology, has been brought to us from observational astronomers and experimental physicists who collect light—very nearly always light, although sometimes there are particles too—from luminous events near the origin of the universe until today. Scientists decode the information subtly infused in the color and intensity and direction and variations of the light across the spectrum in order to reconstruct a detailed map of the universe, a map spread more than 45 billion light-years in all directions and ticking back down nearly 14 billion years in time. In all that is in the vast volume of all that we can see, my favored domain for exploration is the total darkness, the empty space, the vacuity, the great expanse of nothingness, of emptiness, of pure space and time.

Black holes are dark. That's their essence. That's the defining feature that earned them a name. They are dark against a dark sky. They are a shadow against a bright sky. A telescope has never found one unadorned. Since light is the courier for very nearly all of our extrasolar information, bare black holes— those too solitary to tear down sufficient debris—in their obliterating darkness are practically impossible to observe, but not entirely impossible.

We see evidence of black holes destroying neighboring stars. We see evidence of supermassive black holes in the centers of galaxies, their location, dark and unspectacular, marked by orbiting stars. We see evidence of black holes powering jets millions of light-years across, visible in the farthest galaxies in the observable universe. But we have never really *seen* a black hole, which only adds to the thrill of the prospect of hearing them.

There must be black holes out there that we can never see. They are alone. Or they orbit another black hole. Nothing falls into them. Nothing shines bright enough, close enough. We cannot make out the shadow, at least not yet. But if the black holes collide we might hear them ring space and time, sending waves in the curves of spacetime through the universe traveling at the speed of light. If the gravitational observatories succeed and we just marginally make out the reverberations against the noise, we can record the sounds of stars exploding in their final seconds before collapse. We can record the sound of imperfections scraping spacetime as neutrons stars spin. We can record the sounds of neutron stars colliding, possibly forming black holes. And we can record the sounds of black holes colliding to form heavier silent black holes, emitting a billion trillion trillion trillion watts of power in gravitational waves.

As one of the self-described "believers in gravitational radiation," Bell Burnell became hooked with the discovery of the Hulse-Taylor pulsar. Russell Alan Hulse and Joseph Hooton Taylor, Jr., were awarded the 1993 Nobel Prize in Physics for their measurement that confirmed the existence of gravitational waves, though indirectly, by deduction. Hulse and Taylor observed in great detail over several years the orbit of a system known to catalogs as PSR B1913+16. (The PSR=pulsar. The numbers denote the right ascension and declination, the angular locations, and so fix the object in the sky.) They watched a compact dead star, a neutron star, 21,000 light-years away, send radio pulses toward the earth seventeen times a second. The neutron star is a gargantuan magnet that manages to funnel radiation into a narrow beam and sweep that beam around as it spins like a lighthouse. That is to say, it's a pulsar. By delicately measuring modulations in the pulse, they deduced that the pulsar was in a 7.75 hour orbit with another less conspicuous neutron star. So far, already pretty phenomenal. Then they observed the orbit of the pulsar to decay ever so slightly, one full rotation taking 76.5 microseconds less every year, and deduced that energy must be draining from the orbit to cause the dissipation.

The energy loss is precisely as predicted by Einstein's theory of gravity. The orbiting neutron stars drag the curves in spacetime around with them and pump energy into waves in the spacetime geometry. Or, more plainly, the lost energy is carried off in gravitational waves, in the sound of spacetime. Theory and experiment fit together snugly in this fortunate observation.

In about 300 million years the pair will lose enough energy to gravitational waves to fall together and collide, and the final hours would in principle be detectable by an observatory like LIGO if humans are still here and still operating ground-based astronomical observatories, which is ridiculously unlikely for all kinds of reasons. But until then, until the final hour, the gravitational waves are too weak to measure here on Earth. We have no ambition to hear the Hulse-Taylor pulsar. We are in pursuit of others like it, colliding combinations of neutron stars and black holes, pairs in the final minutes of their life together, when the noise is loud enough that we can pick up the sound with our machines a few hundred million light-years away or more. We can see neutron stars in our own galaxy, but they are intrinsically too faint to observe at a distance of millions of light-years. The Hulse-Taylor pulsar, to compare, is only 21,000 light-years away, within the limits of our Milky Way galaxy. So astronomers with light-collecting telescopes can't take pictures of most of the compact pairs before they collide. We'll have to hear them first.

We cannot claim that the gravitational waves from the Hulse-Taylor pulsar were directly witnessed carrying away the energy. We can only claim that the predictions for the gradual decay of the orbit are so perfectly explanatory that by straightforward deduction, gravitational waves must have carted away the energy. Presumably. It's a very good bet. A billion-dollar sure thing.

9

Weber and Trimble

While Joe Weber was alone in the woods at his devolved facility, LIGO secured a significant financial commitment from his previous backer, the National Science Foundation. While Joe maintained his bar instrument DIY style, Caltech and MIT organized a base camp, assembled the essentials, and strategized for a long campaign. While Joe collected the twentieth year of data, evidence of his scorned achievement, journal articles fanned atop a metal cabinet bragged of a new experimental era in which he had no part.

Kip knew Joe in the mid-1960s before Joe's controversial result was announced. John Wheeler took an interest in Joe, and that's how Kip came to know him too. Joe wasn't yet cantankerous. They would hike together in the Alps. They were, at some level, friends.

I ask Kip, "Was he argumentative?"

Kip laughs. "No, because nobody was arguing with him."

"I sensed some jealousy on his part," I say.

Kip concedes, "Oh yes, there was a lot of jealousy. That was a real problem."

I reply, "There was such paranoia and also real complaints all folded together."

Kip agrees. "That's how these things always are. All tangled together."

By the time Kip recorded his conversation in Joe's office, 1982, Joe would have known that Caltech was moving forward with the new technology of ifos. He was going to get beat out again. Kip says, "The saddest part is that Joe was well respected for what he did do and he never seemed to know that."

I listen to Kip's interviews in the Caltech archives, located in an overly confident building, incongruously on a subterranean floor dominated by laboratories. The initial click of the tape recorder sweeps up a theatrical crackle that organizes into static. I surmise that they are recording in Joe's office at the University of Maryland, an office described in press clippings as a warren of paper stacks. I can imagine the room, the entropic décor, the standard metal file cabinets and stacks of curling paper. I'm straining to hear Joe, who may or may not be milling around the room. Joe says it's his brother's birthday on that day, July 20, 1982, the day of the interview, Joe being the youngest of four boys. And in that way that all people enter an exchange with an opening of conventional niceties, they begin.

Joe forms the story of his professional history. He sounds like an innocent man in a criminal interview, depleted by multiple identical criminal interviews, recounting by rote the story he's told countless times to real audiences or just to himself that lead to his alleged crimes. The narrative is tagged occasionally with explanations for how he got here. Kip receives Joe's evidence accrued over the years, evidence to be assembled for his defense against charges Kip hasn't raised. For nearly an hour Kip accepts verification of dates, citations from proposals, selections from publications, detailed expositions of technical specifics, corroboration of Joe's case. The tape is stopped and started a few times as they dig around the office in search of documents.

Gingerly, Kip says to the most controversial figure of his professional life, "Going back early, in terms of you mentioned you got into the gravitational wave business in part because that was a noncontroversial field. . . ." Maybe the subject digs up some rancorous feeling that pushed out over the damaged mantle of his disappointment. In his disappointment, Joe was deflated. But his acrimony is motivating.

Joe says, "The journal *Science,* May 15, 1981, an entire page was devoted to denunciation of me and to proving that Garwin was a much better, a much greater scientist. He really did the important work on gravitational radiation and he demolished everything that I ever did. . . . The important fact is the physics I'm doing right now is the most exciting physics I've ever done in my entire life. . . . It's not advertised primarily because I want to insulate myself from abuse. . . . In fact, I think in terms of getting the vultures by dangling a piece of red meat in front of them, dropping it, and running like hell and undertaking some other field or endeavor, but it's . . . it's . . . all the unpleasantness, it certainly hasn't affected my health, but it's really unfortunate. It does affect members of my family and I think that's incredibly unjust."

Joe took that article around to a family attorney, who suggested he sue. He was advised he could get a $10 million allowance in a payoff for the libel, but it would take at least five dedicated years that Joe was not prepared to spend in court. "It's a question of what you want to do with your life," Joe says.

Joe presses on: "At one point my chairman here gave me two

weeks to clear out. And at another point, Dick Garwin wrote a letter to the University of California administration"—Weber held a part-time position at UC Irvine to be near his wife, Professor Trimble—"and the University of California vice president called me in and said I could be fired on two weeks' notice. He had the letter and I could be fired on two weeks' notice. Eventually I was, I wasn't thrown out of either place. But gee whiz.

"I simply cannot understand the vehemence and the professional jealousy, and why each guy has to feel that he has to cut off a pound of my flesh. Ah, it's a waste of effort. I'm in good health. Boltzmann committed suicide with this sort of treatment. But I just, I don't have suicidal tendencies, I just, ah, keep wondering what the point of it all is."

Kip: "Well, Joe, I really do try and have tried to keep the record straight and to do my part in the history [Joe interrupts with either dismissive or grateful noises, unclear]. I have enormous respect for your contributions to getting . . . for your starting the field and starting it in the direction that everybody else is still following, and that speaks for itself."

Joe: "Still, every now and then people have to roll up their sleeves and stand up and be counted. . . . They probably won't fire me. . . . If you do science the principle reason to do it is because you enjoy it and if you don't enjoy it you shouldn't do it, and I enjoy it. And I must say I'm enjoying it."

Kip: "I agree with that philosophy completely."

Joe: "That's the best you can do."

Joe gets more upbeat, having aired his grievances, and offers Kip a tour around his laboratory and starts to list what they have to see there. "Well anyhow, I'd be happy to take you on a turnaround. Let me tell you what we have. We have a fully operational . . ."

Tape ends. I replace the outmoded technology in its out-moded case and return the contraption to the Caltech archi-vist. I leave the wooden library for the harsh esophagus of the laboratory maze and take an elevator of residential polish up and out of the basement. I head into the unwavering Pasadena warmth and follow the path to my meeting with Kip.

Before Kip leaves, I ask, "Do you know about that letter from Freeman Dyson?"

"Which letter?"

"Oh, it's terrible. Freeman feeling responsible for encourag-ing him, he writes Joe a letter imploring him to stand down."

Kip laughs, clearly stunned, and says, "Wow, he must be a very . . . ah . . . a very optimistic person."

Here is the letter:

Dear Joe,

I have been watching with fear and anguish the ruin of our hopes. I feel a considerable personal responsibility for having advised you in the past to "stick your neck out." Now I still consider you a great man unkindly treated by fate, and I am anxious to save whatever can be saved. So I offer my advice again for what it is worth.

A great man is not afraid to admit publicly that he has made a mistake and has changed his mind. I know you are a man of integrity. You are strong enough to admit that you are wrong. If you do this, your enemies will rejoice but your friends will rejoice even more. You will save yourself as a scientist, and you will find that those whose respect is worth having will respect you for it.

I write now briefly because long explanations will
not make the message clearer. Whatever you decide, I
will not turn my back on you.

With all good wishes,

Yours ever

Freeman

[June 5, 1975]

In winter of 2000, Joe Weber slipped on the ice in front of
the gravity research building in Maryland. To maintain an otherwise unmanned observatory, Joe was accustomed to manual
labor an eighty-one-year-old man shouldn't be. He parked his
car on the top of the hill because he didn't think he'd get it
back up and out again and aimed to cover the rest of the jaunt
on foot. It was two days before he was found with some broken
bones and an incursion in his chest that allowed lymphoma to
take hold. The bones and the lungs never completely healed.
Eight months later, on September 30, his wife, Virginia Trimble, received a call from the hospital in the middle of the night,
just in time to write his obituary for that morning's deadline of
the *Bulletin of the American Astronomical Society*.

Virginia Trimble made the decision long before her husband's death not to spend her energy in defense of his record.
"Science is a self-correcting process, but not necessarily in one's
own lifetime," she tells me rightly when we meet in the computer room on the UC Irvine campus. Her attitude now, and his
then, she says, was that Weber detected something; whether or
not they were actual gravitational waves was unclear. The fault
in all of the disputes was that there had never been any other
"carbon copy," an exact replica of his instrument run over
the same times, a true hard-line, literally oranges-to-oranges
comparison.

"Joe's own point of view was that no one had ever quite repeated what he did, so that nonconfirmation claims were not entirely honest. The two groups that did the most similar things—one in Japan, and one in Rome under the late Edoardo Amaldi—did see events like the Maryland ones. Indeed there are a couple of papers reporting coincidences between Rome and Maryland at the time of SN 1987A [a supernova explosion close enough to be seen by the naked eye in 1987]. At the very beginning, before the July 1971 Copenhagen meeting, Vladimir Braginsky sent Joe a postcard saying he had confirmed the results. Braginsky didn't get an exit visa and was not in Copenhagen. Well, he *could* have meant 'repeated the experiment.' Of course he changed his mind later in any case. I tried to show the postcard in my talk at the last Texas symposium in São Paulo in December 2012, but the projection facilities turned it upside down and backwards."

Virginia points me to a scan of the original, a cheerful winter holiday postcard, a collage of stamps on the back where Braginsky wrote,

Dear Prof. Weber

I wish you a happy New Year. I hope to see you
in Denmark for to say that I have confirmed your
experiments.

very truly yours

And it looks as though he signed the card "Braginsky" in script. "Confirmed your experiments," could well mean "repeated your experiments," as Virginia acknowledges. (Braginsky soon published negative results, which inspired "quite a severe fight" with Joe Weber.) I had trouble identifying the date on the postcard, but Virginia places the Denmark meeting in context:

"Nineteen seventy-one," she says. "This was before I had met either of them."

For historical accuracy, Kip remembers Braginsky's presence at the conference vividly and insists he did get an exit visa. There was even a dramatic altercation in which anti-Soviet speeches drove the entire Soviet delegation to march out of the lecture hall. At Kip's pleading, Braginsky returned to make a conciliatory speech for which he was later denunciated.

She called him Weber and he called her Trimble. They married in March 1972 after a cumulative three weekends together. She laughs. "Weber never had any trouble making up his mind." Twenty-three years her senior, he always insisted she do what she wanted and needed to do. Perhaps trained in part by his first wife, Anita, a physicist who took a protracted break to raise their four boys, the widower had no reservations about Virginia's work, her independence, or her IQ. (Stratospheric. In an issue of *Life* magazine with a now-vintage cover, in an article titled "Behind a Lovely Face, a 180 I.Q." about the then eighteen-year-old astrophysics major, she is quoted as classifying the men she dates into three types: "Guys who are smarter than I am, and I've found one or two. Guys who think they are—they're legion. And those who don't care.")

Disappointed about the downturn of a relationship, Virginia vowed to marry the next man who asked. She did give her previous suitor a chance to intervene. She wrote him a letter that stated, "I'm going to marry Joe Weber, if you want to stop me, call me in California." She didn't know he had gone to Princeton, another disappointment, as they had each promised to tell the other of his or her travel plans, and he never received the letter. Or so she hopes. Although still friends, she hasn't the heart to ask.

She married Joe as planned. "We both felt we got the best of the deal. We agreed on many things like the benefits of getting up early and a good breakfast. Also when a man shows up to a restaurant with a trophy wife, he always gets a table."

Joe left out some candles and matches one evening to see if she would know what to do and discovered she could sing unknown harmonies to the Kaddish. She was interviewed by learned men, did a *mikvah* with some women or something (I didn't quite follow, being poorly informed about tradition), passed an oral test, and became Jewish. And though both Joe and Virginia were proclaimed atheists, Joe was proud to let his sisters know he married a good Jew. She maintains the tradition to this day if not the literal beliefs, and on the day of our conversation, she describes a singing engagement in synagogue that she is looking forward to.

They were both driven to make a living. Joe's father was a carpenter and a good union man who wouldn't consider a non-union job. When the Webers found their furniture on the lawn, they realized he had defaulted on his mortgage. Virginia says that if she came home and saw her father's car in the drive she knew he had been fired again. "He was a rather good chemist but a rather poor businessman."

"Barriers fell away in front of me," she offers. Virginia suggests her independent funding from the Woodrow Wilson Foundation eased her admission to Caltech. Her advisor, the famous astronomer George Abell, nominated her for the scholarship traditionally reserved for the humanities, and she stood out despite her scientific inklings for her knowledge of hieroglyphs and archeology. While at Caltech she posed (nude?) for Feynman's drawing class. She also brought in pennies, she says, doing voice-overs and commercials. She was "Miss Twilight Zone" and toured all the Nielsen cities to boost ratings. She was also a fine astronomer. Women were not allowed to observe at Mount Palomar until Vera Rubin broke that wall a year ahead of her. Another barrier felled, Virginia Trimble observed at Palomar. In her third year, having demonstrated her tenacity—particularly manifest in the fact that she still hadn't married, she suspects—she was awarded a fellowship from the National Science Foundation. When she arrived at Caltech, she was delighted. "I thought, 'Look at all of these lovely men.'" In her seventies, with her coral dress, matching shoes and lip color, Moon earrings, and gold animal-head ring, she beams. Still a lovely face. And still an IQ of 180.

"Well, we're both a little Aspergic, both a little odd I suppose," she says surprisingly frankly. "He used to say, 'The best thing that ever happened to me was marrying Virginia.' I never had a problem asking him for help. When I fell and broke my hip last September, I spent four days on the floor of my apartment singing songs and reciting poetry until I was found. I didn't want to think of Joe's two days on the ice. I hate the cold. But I did think, 'This wouldn't have happened if Joe was here.'"

"Did Joe ever consider joining the LIGO collaboration?"

"He wasn't asked, and I don't know what he would have said."

"Was he frustrated by LIGO's funding successes?"

"Not frustrated. He felt ill done by. He was always a fairly cheerful person. If that hadn't been instantly obvious, I would not have married him. People who worked for him found him very attractive. He treated them well. Secretaries loved him.

"He worked his unpleasant experiences into anecdotes. When he survived the sinking of an aircraft carrier in World War II, he had his closest death experience of the war. A monkey on shore threw coconuts at him.

"He would say he invented three major experimental fields:

quantum electronics, gravitational radiation, and coherent neutrino detection." (For the record, let me jump in here and say the first is an indisputably important and verified scientific discipline, one for which he might have been included as a recipient of the Nobel had credit been doled out differently. The second is still controversial, and the third might even be hopelessly controversial.) Then Virginia adds the uncontroversial claim for which Weber will be most remembered: "Weber's goal was to bring the Einstein equations into the lab. He felt, and I think it's fair to say, he succeeded in doing that."

10

LHO

The initial LIGO detectors built around 2000 did not hear any cosmic sounds. That generation of instruments proved that the technological feat was possible, but they were not actually sensitive enough for a first detection. Or maybe nothing is there to hear. We put aside misgivings. We're at the summit already, the surface of the Earth. The summit is a location, wherever we are. It's also a time, in our future, when the advanced machine will be fully operational. On the ascent we lost Weber and, for all intents and purposes, Ron Drever. Still, the numbers on the climb grow. No matter who falls away, others take their place, and the ascent continues. The expedition is alive; the march picks up pace and heads toward the collision.

The LIGO laboratory in Washington, also known as LHO for LIGO Hanford Observatory, is located on a remote square of government-owned land in southeast Washington, home to the Hanford Site—the venue for the world's first nuclear reactors. John Wheeler spent the final year of the war at Hanford, having designed the reactors. Plutonium separation facilities extracted the radioactive element for Fat Man, the second and last atomic bomb (to date) dropped from a plane onto a city. What began as a sparsely populated area became veritably remote in 1943 when part of the classified Manhattan Project moved in and the War Department summarily moved out (evicted) residents within nearly 600 square miles. The Cold War encouraged expansion of the nuclear facilities for the purpose of amassing threatening stockpiles of weapons. Since the late 1980s the Hanford Site ramped down as a nuclear production center and became a nuclear cleanup facility. You have to do something with those unfortunate excretions, preferably diverting your intentions away from the Columbia River.

So there it was, an unused and unusable remote location in the not quite desert—it's actually known as a shrub-steppe, which while rain deprived has more cover than an official desert. Tracks of perennial shrubs give the impression of a poorly laid field, a messy handwoven farm. The flat stretch is nearly without perspective until the reactors emerge on the horizon topped by cooling plumes—the unintimidating cousin of the atomic mushroom cloud—leaking into the cumulus, smearing the innocuous weather clouds' cartoon poses.

Several miles distant are a few small buildings that constitute the bespoke LIGO lab. The buildings are new and planar and almost white, an architectural contrast to the round-bellied reactors that mark the periphery from the LIGO point of view. The grounds are manicured, with green crisply groomed bushes planted so sparingly among nonindigenous pebbles that the effect is of a half-finished but carefully rendered diorama. Almost intentionally artificial.

I get there early enough to join the daily 8:30 AM meeting in the control room. Michael Landry, in charge of installation of Advanced LIGO at LHO, casually listens to status reports delivered from around the room. About twenty people lounge among the tables and chairs, slightly outnumbered by two columns and three rows of computer monitors. One guy bounces on one of those giant exercise balls. The meeting is efficient and brisk and concludes with Mike: "Let's work safely today. We're done." The delivery is earnest yet habitual.

People tend to hang out in the control room during normal working hours, coming in and out of the lab wearing doctors' scrubs, but it seems to me a darker blue variant, the new regulation, an additional layer of protection against contaminants. I'm not inclined to press the analogy of the doctor and patient; it doesn't really hold all that very well except the six monitors on one wall and the seven on the opposite do display all kinds of metrics on the machine and do sound an alarm if it's in jeopardy. There are multitudinous cameras and sensors at many points along the anatomy of the detector. The control room is never empty. Operators are on duty 24/7, each on an eight-hour shift. During science runs, the machine has to stay in lock, which means the mirrors are maintained in a narrow range around a fixed separation. A complex feedback loop adjusts to restore the mirrors to their original separation as they move, kind of the way a thermostat tries to keep a room at a selected temperature. The instrument measures the tiny displacements around this locked state and keeps track of the efforts to restore the mirrors' locations. If the machine goes out of lock, an alarm will sound and a screen will flash yellow or red. Sometimes people change the alarm sounds for a gag.

Everyone sort of smirks about the semiautomated aspect of the control system. Control room operators confess it's all a bit of a black art. Not just like a knock on the side of the head of a temperamental television. Darker and more mysterious. *A Black Art*. Lots of people intone this phrase to me, breaking eye contact, offering the smirk instead. I think they mean it. They don't want to alarm me, to draw attention to the challenge. It's not always obvious how to keep the machine alive. It takes months to learn to operate the controls through the graphical user interfaces (GUIs). When Advanced LIGO is fully assembled, the number of readout channels will bloat to 200,000, the number of control loops to 350, and then what? Who will be able to control them all, keep the laser flowing down the veins?

If it's a windy day or dump trucks are abundant on the roads off the Hanford Site, then the thing might not get back into lock at all. The nights are easier to maintain, and from the demeanor of some of the operators I spoke to, hired locally, not usually drawn from the pool of academics, the nights aren't a bad time all around, just them in the empty control room big enough for a hundred.

There are about twenty people milling around the control room—people in their dark blue scrubs, tapping their own heads with wrenches and screwdrivers thoughtfully or self-punishingly. Okay, they do actually look like doctors, especially when there are lots of them as they consider the readouts on the monitors and discuss the delicate condition of the patient.

There's definitely camaraderie encouraged by the remote location and the relatively obscure task. Arranged with their backs to the open door, people face the diagnostic screens and the unusually large digital clocks, one recording local time, the other GMT. They lob jokes about the desks and shout questions to the air for anyone with the attention available to answer. "How do you spell 'defunct'?" I shout back "C" as I stray out of the room and head for the lab.

I'm aware that the control room shares a wall with an air hangar of a laboratory, the Laser and Vacuum Equipment Area—the LVEA. While the Caltech 40 meter prototype is hidden by a trailer, this full-scale observatory cannot be contained in a building. The LVEA is roughly 30,000 square feet and houses only the apex of the ifo. Just beyond the wall, two beam tubes, each 1.2 meters in diameter and 4 kilometers long, punch right through the laboratory out into the northwestern shrub-steppe. The stainless steel tubes are merely 3 millimeters thick (with stiffening rings to support the structure). Unrolled from spools and then spiraled into cylinders, welded together in 60 meter sections to span the full 4 kilometers. Protective cement encasements surround the tubes, paralleled by an access road to smaller laboratories at the end stations of the tunnels.

Two of the largest holes in the Earth's atmosphere are in those LIGO beam tubes, just there beyond the double doors of the control room. There's less stuff in those tubes than there is in the empty space between galaxies, which has a very little bit of stuff known as the intergalactic medium. Two of the biggest holes in the atmosphere and eight times less stuff than parts of outer space. (Although the emptiest regions of outer space are emptier still.)

The scientist-designed vacuum system was both cost-effective and impressive. While there are vacuum chambers on Earth with even less stuff in them, none are as enormous as LIGO's combined. The tubes were brought down to vacuum in 1998 and have not been brought back up to atmospheric pressure since. In the transition to Advanced LIGO everything is being replaced but the nothing. The nothing needs to be kept at nothing for the lifetime of the experiment. If the vacuum is destroyed, that would be the end of it. Mike Landry says, "We'd all go home."

At 3:00 AM one night, a supervisor from the security force at the nuclear facility wandered into the LIGO facility and asked, "Did you guys hear that?" Mike drove down the access road to find a truck crushed into the cement encasing one of the arms. The security force that patrols the Hanford Site is federally empowered, lethally armed, stocked by pretty big fellas with intimidating gear, and at least some officers have a penchant for speeding in the dark despite a gap in their knowledge of the local geography. Storming across the shrub-strewn plain at 50 miles an hour, the officer crashed into an interferometer arm, breaking one of his own. And also a rib.

The crash didn't puncture the vacuum but it could have, and it's conceivable that while a dime-sized puncture might only whistle ominously and destroy the experiment, a big-enough aperture could be lethal, like opening a vent on the Space Station, only the empty space is inside the ship.

Cars are bad news even if they don't crash into the tubes. LIGO is extremely sensitive to seismic vibrations. It is, among other things, a spectacular seismometer. The machine can hear trucks trafficking down the access roads for instance. Even air acoustics has been a problem, and the data analysts found a noise correlation with the arrival and departure schedule of the local airport.

The Sun and the Moon cause the mirrors to sway, and magnets are required to restore the mirrors to their baseline location. There are also seismometers to detect local motions of the Earth with attendant hydraulic systems to compensate for those displacements. All of this constitutes forms of noise, to be distinguished from a bona fide signal. We listen to the raw sounds of the instrument. It whirs with the tidal pull of the celestial bodies, the grumbling of a still-settling Earth, the remnants of heat in the elements, the quantum vibrations and the pressure of the laser.

The mirrors are stunning. To our eyes, they look perfectly transparent, practically invisible. They are terrible reflectors of optical light. All of their muscle is in their ability to reflect the laser light. Manufacture of the mirrors is outsourced to companies very good at such things and then sent around the world for different processes, including eighty layers of coatings to make the best mirrors so highly reflective with almost no losses. At the frequency of the laser, the mirrors are 99.999 percent reflective.

Extraordinarily delicate glass fibers suspend the 42 kilogram mirrors because they cannot be bolted to the tunnel ceiling. If bolted, the mirror would not sway with the changing space, bob with the ocean wave. So there's the rub, an unforgiving tension between stability and sensitivity. The glass fibers are about twice the thickness of a thick human hair, so delicate that they break if touched with malintentions, yet strong as steel when stretched in length.

Head of LHO, Fred Raab, called the entire game a balance of finesse and terror. A megawatt of laser light trapped in the cavity between mirrors constitutes an aggressive accumulation of power. When the machine drops out of lock, that megawatt gets dumped onto a photodiode designed to absorb only a very delicate collection of photons. In one incident the camera got fried. Stainless steel shutters were then designed to close quickly enough to protect the collection apparatus. In another dropped lock, the accrued power was thrown onto the efficient shutters, but even the metal got smoked. Singed material wafted off the shutters into the vacuum.

Another near disaster has been attributed to an earthquake in China that knocked around some of the smaller optics. The system that tries to damp the motion couldn't regain control and kicked the optics back and forth for hours. In an attempt to relock the ifo, an operator inadvertently steered the input beam like a laser cutter across wires, heating them until they melted.

Wires broke. Things fell. Incidents like that only happened a couple of times, and there are earthquake stoppers to catch the mirrors on the seismic shelves just in case. Usually earthquakes around the world are registered to less disastrous effect.

Mike Landry and I get suited up to enter the LVEA and have a closer look at the operation. The entire air hangar is maintained as a class 10,000 clean room, which tolerates about that many dust-sized motes per cubic foot. In comparison, on average New York has at least a million contaminants per cubic foot of the microbe or dust or chemical variety. (On a visit to the Louisiana site I'll watch an hour-long talk on elevated clean room guidelines that includes a hands-on demonstration involving surgical gloves, Jif peanut butter, and a squirt bottle of isopropyl alcohol.) The room is cold (sweat is a contaminant) and vast, with ceilings somewhere between 30 and 40 feet high. Tracks are built into the walls and cranes glide along those rails above us. From down on the ground the rails are branded with their 5 ton capacity in all caps. We wear hard hats.

The chambers have to be sealed off from the tunnels with brawny gate valves so that they can be brought to atmospheric pressure without flooding the tubes with contaminant air. Viewed from a set of temporary stairs and walkways that offer a higher perch, the collection of chambers at the apex became known as the beer garden, which gives you a sense of their appearance—a cross between a brewery keg and an H. G. Wells submarine. The top of the chamber is opened and the entire suspension cartridge is lifted up and in by a professionally operated crane. The chamber is then sealed—well— and pumped back down to the vacuum level of the arms so that the gate valve can be opened.

After an eight-week installation they opened a viewfinder at the end station (this was in Louisiana, not here in Washington) to see a live 2 inch spider tacked to the inside of the glass. Bugs are a problem at both sites. Mice are a problem. "Sorry, buddy," Mike said as he stepped on a spider in the small room at LHO where we try on our clean suits. A few minutes later in the end station lab, a senior experimentalist breaks mid-sentence, eyes intensified above his mask, to take down a moth on the wrong side of the plastic strips that wall in the most precious components. As Mike picks up the dusty carcass off the floor he says again, "Sorry, buddy."

The tubes pierce the walls and bore another few kilometers into the dry terrain. Between the beam tube and the beam tube encasement that connects the LVEA to the end stations 4 kilometers away is an area large enough for a person to walk the length of the instrument, which no one did until Rai Weiss discovered an infestation of mice, wasps, black widows, and snakes. The wasps like to eat black widows, which they put in hexagonal nests and keep anesthetized but alive until the wasps get interested. The black widows make hydrochloric-acid urine, which corrodes and stains the stainless steel, which indeed has visible stains. There are no swimming pools made of stainless steel because it's not stainless to chlorine. "The black widows are interesting," but after a lot of investigating Rai concludes, "the real culprits are the mice."

Rai walked up and down the tunnels to diagnose the problem, which was worse in Louisiana than here in Washington: tiny leaks about one-thirtieth of a hair width thick (caught and sealed) in the vacuum. At the sites, I always hear these stories about Rai. Rai crawled into the tunnel that time. Rai found the broken bits of glass in the tube. Rai dispersed mice and wasps and all manner of vermin. Rai's walking the beam tube again. Rai always under the proverbial hood.

Rai lets me tag along at LHO while he runs a little experiment on the vibrational modes of the tubes. The access road has to be kept clear of problematic tumbleweed. Dried shrubs uproot and roll along the plain, accumulating along the tube encasement like thorny dust motes against a wall. To clear a path through the dry blown weed they are gathered and packed into rectangular bales, like hay bales, which are then distributed just outside the perimeter of the diorama, like craft materials sharing a table with the sculpture, maybe to be used or maybe discarded. I like the tumbleweeds, both the untamed form and the packed rectangular form. They stitch the artificial grounds of the lab to the natural landscape.

Rai offers, "Tell me if the smell is too much. It's far worse in Louisiana. Last year I got fungal pneumonia." The air gets better as we open a few of a total of fourteen doors to the cement encasement that protects one of the arms. The smell isn't that bad, but I'm grateful for the ventilation. "I've always walked the tunnel," Rai says. The beam tube was his scientific responsibility for years. The tube vibrates. He bangs it for me and loudly it moans and sustains. As LIGO pushes down to better sensitivity the experiment becomes aware of all of these low-frequency seismic vibrations, which have always been there but were less important at poorer sensitivities. Rai lets me help set up, like you might let a child, by tightening a vise and holding a cable. Rai does whatever he can to push the project along, despite being officially retired and being in his eighties. He does this work so that others don't have to. Rai is clapping and stomping and walloping the tube.

"This takes a lot of patience," I say, not proud of the obviousness of the statement. "Are you patient?" I ask, undeterred. "No, and neither are you," he says. "You can tell?" "Yes. You keep finishing my sentences," he says, not unpleasantly. I'm chagrined. He waves away any criticism: "It's okay. It's good."

After we clip some cables together and attach a small instrument to one of the spines around the tube, I have to sit in the car while Rai performs diagnostic measurements of the beam tube vibrations. The car is getting hot because I forgot to open the window even after he told me to, and now the desert sun is cooking. Any noise I make could ruin Rai's diagnostic test. I imagine the clunk of a car door handle and possible squeak of the opening door and decide to broil quietly in the blanched daylight.

On our drive off the Hanford Site that evening and over the several days we spend together, Rai recollects the early days of LIGO and the Troika in the 1980s. The management structure was hopeless, he tells me. Ron Drever has been described as constitutionally incapable of sharing power, or of trusting other people's judgment, including Rai's. The Caltech team had to hang together, despite Ron, for the project to succeed. Rai says, "I would do everything necessary so that the thing could keep going forward. Whatever it was.

"Ron was very difficult to deal with. At the time, I had a lot of respect for Ron, of a different kind. I began to understand Ron better as a scientist. I also found out why he was impossible to deal with: He doesn't think the way you or I think; he thinks in pictures. And he doesn't remember what he thought the day before, so you could never make a decision. You could watch his process. He'd go through the same logic about a decision on how big the laser beam should be or how many mirrors there should be—I don't know, pick anything in the interferometer. And you would discuss this with him, and you would get to the same point, and he would agree that his viewpoint was not right—or maybe he wouldn't quite agree—and then the conversation would start all over again the next morning, from exactly the same place. And we'd come to the same conclusion. This would go on day after day; you'd never come to a resolution. That was one of the troubles.

"Ron would effectively tell [Kip], 'Look, you've gotten me here under false pretenses. I thought I was going to get this and this and this, and now look where I am. I'm in this terrible situation with Weiss and these MIT people who are going to eat me up,' and all that sort of stuff. And Kip, I know, felt terrible underneath about that, because there was a little truth in it. I mean, I don't think that Ron ever thought he would have to deal with somebody else."

Drever was again playing Mozart, and Rai worried in his darkest moments, his self-esteem faltering, that he was being cast in the role of Salieri. Rai had his own ideas about the instrument, about design implementation. But he had to put his own ego aside, not without some personal suffering, to get the thing done. He worked on site selection, or an industry study, tested mirror coatings, built his own laser. Even now he'll work wherever on whatever is needed whenever, sweeping away wasps, walking the tunnels, testing systems, building electronics. I cannot tell you how many times I have heard someone say, "We better ask Rai."

"So Kip was still in there; I mean, he had to." Kip tried to hold them together, to balance the egos and the authority so that the peculiar combination of personalities might combine effectively. He recommended different domains of authority, gave them equally important titles, like chief scientist in charge of this or chief scientist of that. Kip's arbitration was facilitated by his unflappable temperament and by his personal computer. He was the only one with such a thing. The Troika could send a half-baked idea into the aluminum casing, transmitted by Kip's typing, and out would come a decree, printed in black on white. A resolution became more official after its transformation through that computer, a maker of authority. But the real decisions were never tapped into the keyboard, never materialized in print, essentially were never made. The tensions between Rai and Ron, the incompatible styles—Rai's gumption and determination to move forward, the dreamy pictorial nature of Ron's cleverness—hampered any effectiveness each might have had without the other. Ultimately, there was no decision that could be made by the three of them. "None," Rai says.

"This is an exaggeration," Kip later corrects, "but not by a lot."

Rai says, "The watershed in the whole thing came when Dick Garwin wrote a letter to [the NSF]. That's leaving out a lot of history, but let's leave it go. That was in May of eighty-six. This is now three years into the Troika.

"Garwin wrote a letter to NSF. Maybe he felt he had killed this field. Now what were we doing, resurrecting this. At Garwin's suggestion, NSF demanded a summer study. . . . And so they called me and I think that rankled people at Caltech somewhat. Of the Troika, they demanded of me that I run the summer study. After they put all that money into the industrial study of the Blue Book. Okay, I think it was legitimate."

Garwin was a very influential IBM scientist and among those to build Weber bars after the infamous claims of discovery in 1969. His opinion mattered, having served as advisor at high levels. He played a role in halting the Star Wars insanity as well as potentially disastrous industrial escalations, like the plans for supersonic airplanes in the 1960s that would have cruised in the stratosphere, getting passengers from New York to California in a fraction of the current travel time while poisoning the delicate atmospheric layer irreparably. Garwin had annihilated Weber. He considered the slaying a public service. He was not pleased to learn of the resurrection of gravitational-wave experimentation, and at such a staggering cost.

Rai continues, "Okay. Well, Dick thought he had slain this dragon, and then all of a sudden out of the ashes here was a phoenix that had risen.

"The point is, what it did was it brought out the problems in the collaboration but also brought out that there was a lot of technology already developed. I got people in the laser business, in the precision measurement business, some of the bar people, people who had measured things so beautifully. We had a meeting with all the subjects that went into this, but the thing we couldn't discuss properly was management.

"And I told them what the problem was. I said, 'Look, this thing is dead unless you make a recommendation that there be a single director. You've got to get rid of this Troika. It doesn't work.' It turns out both Kip and I without telling each other used that meeting as a way of telling that committee the management sucked."

Kip emphasizes, "This November 1986 meeting was tremendously important . . . it got us a ringing endorsement in all respects except management." The report served as an in-depth review process that encouraged them to advance with a construction phase to expedite instrumentation development. The positive assessment gave the NSF's Isaacson the confidence that the project could move forward with a design and construction proposal (two previous proposals submitted by the Troika had been rejected). There was one condition: They find a single director. All members of the review committee signed off on the report, including Garwin.

Rai says, "And what happened out of this was Robbie Vogt, then Caltech's provost. Okay?

"Robbie did some good in the beginning. I hate to say it.

I'm going to be very fair about this; I think I have to be. The first thing I did when I heard about Robbie is I started calling around the country and got tremendous recommendations for him. He had done wonderful things. And only one guy was totally honest with me, and I didn't believe him. He said—I'll never forget the words—'Well, you and Ron won't be the same after he comes in.' And I didn't know what he meant. I remember asking him directly, 'Will he screw it up?' And he said, 'Oh, no, no, no. He'll make it work. He'll make it happen. But you and Ron will not be the same.'"

11

Skunkworks

Rochus E. Vogt had been fired as Caltech's provost, which might not sound like a very good recommendation for director of a highly unusual, technologically cryptic, titanic nascent project. Let's not read too much into it, but Vogt was a title given to a manager who presided over certain territories in the Holy Roman Empire. In other words, Vogt sort of means "provost."

Despite the prophetic name, Robbie says of himself, "I am well known as a person who detests any authority."

As provost he expressed an allegiance to Caltech that exceeded any to a specific country, and while he didn't love the phrase, he concedes "hired gun" was an accurate portrayal of the provost's errand. His loyalty to intellectual institution over country might be partly defensive. German nationals who grew up in parallel with Nazism are better off with a personal story that puts them at odds with the rising authority, or implicit collusion is the fallback scenario and would make for an awkward provost's bio. For the record and on record, he had all of the right political reactions to totalitarianism (horror and rejection) and all the right political reactions to the Constitution and the protection of individual rights (admiration and acceptance). But Vogt's staunch fealty toward Caltech was a good alternative to any nationalism.

When I meet him in his Caltech office he says, "Yesterday was the eighth of May. On the eighth of May 1945, I was fifteen years old. I had just been a prisoner of war, and I swore to myself never in my life will any stupid authority have any power over me."

I knew going into the conversation that the Nazis had dismantled his privileged upbringing in southern Germany. After the war, he was demoted to a farmhand and then steel mill worker. Eventually his studies brought him to the more prosperous United States, and by then he had already become Robbie, nicknamed by an American soldier turned unlikely friend. The American was effectively a weapons inspector assigned to his German university to ensure there would be no manufacturing of nuclear weapons, and the German steel industry engineer Rochus, in his capacity as student representative, was liaison. None of this explains why Vogt got fired.

Vogt was the principal investigator on one of the primary experiments on the Voyager mission, the Cosmic Ray System. Currently the two Voyager spacecraft are more than 15 billion kilometers from the Earth, hurled farther than any other human-made objects. They are very nearly between stars, interstellar, shrugging off the Sun's magnetic cloak, outer steel exposed and brushed by the winds of more distant stars. A little dramatic, but true. Vogt fought to extend the mission objectives to interstellar space. He argued for the spacecraft to carry more hydrazine—an unpleasant chemical needed to orient the spacecraft beyond the solar system—which took payload away from the planetary scientists. He explains, "The farther we get away we have to reduce the bitrate in order for it to transmit. . . . The plutonium generators, which provide the power, will be good for another five to ten years and then it's over. Then there's not enough power to communicate. . . . Easily another five years when we will be in interstellar space measuring the galactic cosmic-ray spectrum. Except when I say 'we' not 'I.' . . . Now they make the discoveries. And that is the only thing I regret. Administration has deprived me of that. And that hurts. But only because it would have been fun to see first."

Launched in 1977 from a crusty hunk of Earth, Voyager is unmanned but etched with messages about this place, the portrait curated by a committee chaired by Carl Sagan. The mission's least serious purpose: to serve as glorious bottle bobbing on interstellar winds protecting a collection of souvenirs in the event that another living creature should be out there and interested in the mission's inventors. Some citizens objected, since the etched gold phonograph records revealed in diagrammatic form our delicate planet's location to potential aggressors. But the extraterrestrials will have to find Voyager first, a minuscule metal bit in the rattling emptiness of interstellar space. In tens of thousands of years the spacecraft will still not have stumbled into another star system. And finding us by the conventional methods of a galactic explorer, whatever these might be, must be easier than finding Voyager first and deciphering its message only to turn around and spot our solar system in peripheral vision.

To be provost, Vogt relinquished to others leadership on his mission before Voyager broke into the outer reaches of the Sun's magnetic influence, when the real spoils for the Cosmic Ray System were to be garnered. On accepting the provost's office, he mused (why even consider the possibility?) over whether he could return to cosmic ray experimentation if he were fired.

In an interview early in his term, Vogt foreshadowed. "If I went back in, my colleagues would have to feel sorry for me, because I'm out of touch. It would be bad to embarrass people that way. So it's clear I will have to enter a totally new area." Murph Goldberger, then president of Caltech, did have Vogt fired within a few years. If Murph could have simply fired him, if executive power had been his, he probably would have swung the axe sooner. But the board of trustees must be complicit to dismiss a provost, and although Vogt was considered adept and even inspired as an administrator, and much liked by the trustees, he was also portrayed as paranoid and difficult, and maybe the portrayal didn't feel that implausible or unfair. Resentment and blame blistered the administrators' relationship, and their union had to be dissolved. This bit gets gossipy and maybe isn't that interesting or relevant except to position Vogt right where fate needed him.

Negative forces pushed him to that nexus—effectively unemployed (though not unsalaried), unable to return to his previous scientific discipline ("It would be bad to embarrass people that way"), housed dispiritingly close to the men's restroom in the basement of a physics building (having no laboratory or group), recalibrated by disappointment (why didn't the faculty rise up in solidarity when he was sacked?), ready "to enter a totally new area." Positive forces pulled him with equal magnitude—ambition, vision, vitality. All that was needed was the pressure differential of the intentional collapse of the Troika, practically right underneath him, just down the literal and proverbial hallway (actually, after he was fired, his office was one floor below Kip's), and Robbie Vogt was sucked in to position.

He never wanted the job. And he would be fired as LIGO's director too. "I've had no contact with the LIGO project in twenty-five years," he warns as though nothing could come of our conversation. But he welcomes me into his large corner office in a building that can only be defined as LIGO headquarters, at the end of the corridor that accommodates colleagues he hasn't spoken to in nearly a quarter century. Vital scientists on the LIGO team have seen but never met him and express disbelief, concern even, that I will be down the hall in the famed corner location with the notorious, sizable, formidable Robbie Vogt, as though he haunted a dark and terrifying childhood closet that was better forever sealed.

The day Vogt's term as provost ended, the chairman for Caltech's Division of Physics, Mathematics, and Astronomy came to his office and Vogt said, "Take your list away," referring to any departmental business that might need a provost's eye. "It's over. I just resigned." The division chair, Ed Stone, responded, "Oh God, that's terrible." Vogt explains: A search committee had recommended him as the director of LIGO. Stone's charge was to approach him when he was in a good humor, to find a flattering light in which to offer the job. But given his sacking as provost, the timing congealed the offer into a consolation prize. Vogt's reply: "Ed, you're out of your mind. I won't touch it."

Kip speculates that while Ed Stone may have tried to gauge Robbie's response as a serious candidate for director on that day, the job was not offered to Robbie until several weeks after he "stepped down" as provost.

As we settle in his office, Robbie tells me he was coerced into being LIGO director. "I refused and my refusal was not accepted." His resistance was based on a general suspicion of the subject due to Weber's resonant bar detectors and his disputed claims. "Incidentally, Weber was a tragic figure. He was a good scientist actually, but he was so obsessed with detecting gravitational waves that he grossly misinterpreted the data."

Eventually Vogt conceded to considerable administrative pressure. (He was threatened, he says.) "But in the moment where I decided to take the job it was my project and I was totally dedicated to it. And I needed that dedication."

By 1987, Vogt was the director of LIGO and a new domain was his. The Troika—Ron Drever, Rai Weiss, and Kip Thorne—were suddenly free to pursue their destinies within the project. Robbie has only praise for Kip, "who deserves the Nobel Prize," and for Rai Weiss, "a good scientist. A good man." He even praises Ron Drever: "I knew that Drever was a very brilliant scientist. He was just a lunatic personally." (For the record, the consensus on the street is that the Troika as a group will be under consideration for the Nobel.) Robbie brought all of his commendable characteristics to the role and all of his flaws. Someone observed, and it was relayed to me as an apt description thirdhand so I leave this without attribution and to your discretion: "No one was more insightful and creative than Robbie, no one better at solving a problem. And no one was better at creating one."

In 1989, Rochus E. Vogt, as principal investigator, submitted to the National Science Foundation the culmination of the efforts of the joint Caltech-MIT team—an in-depth, thoughtful, 229-page proposal titled "The Construction, Operation, and Supporting Research and Development of a Laser Interferometer Gravitational-Wave Observatory." The proposal opens with the quote:

There is nothing more difficult to take in hand, more perilous to conduct, or more uncertain in its success, than to take the lead in the introduction of a new order of things.

—MACHIAVELLI, *The Prince* (1513)

Rai calls it a masterpiece. Everyone on the project was thrown at the task and what emerged was a thorough, defensible, compelling conception of LIGO, two 4 kilometer observatories working in unison on different U.S. coasts. Rai's haiku was ultimately thrown down to NSF for $193,918,509 in a lucid case for a viable instrument—a new portal to the universe—to be built over four years beginning in 1990. "The LIGO," which has since lost the "the," lays out in the summary of the document its two objectives: "(1) Tests of the General Theory of Relativity . . . and (2) the opening of an observational window on the universe that differs fundamentally from that provided by electromagnetic or particle astronomy." With that proposal, Vogt initiated, if not fulfilled, his destiny as director of LIGO. And the NSF approved the money.

Two hundred million dollars don't just land in a bank account. As significant as the sum sounds, the budget was not staggering in comparative terms, given the budgets for particle accelerators, for instance, which reach into the billions. Still, LIGO was the largest endeavor undertaken by the National Science Foundation, and a special allocation of funds had to be requested from Congress. A major hurdle had been cleared, for sure, but there would be more. The recommendation of funding from the foundation began a long battle to encourage congressional approval. There were congressmen who had LIGO on their targets because, according to Robbie, they believed the project (and maybe science in general) to be a waste of money. Congress stalled delivery of the funds and thereby construction of the sites. For two years, Robbie commuted to Washington to woo the Senate and the House. He became a well-known figure in congressional halls, among chief clerks, and in appropriations committees.

Robbie convinced Caltech he needed a lobbyist, a move unpopular among some of the Caltech faculty to this day. After much resistance, a professional was brought in to advise him, and he went back to D.C. better equipped to unplug the clog. Robbie, though well prepared for the March 13, 1991, House of Representatives' hearing of the Committee on Science, Space, and Technology, was entirely unprepared for the counter testimony. It was at this congressional hearing that the reputable astronomer Tony Tyson delivered a damning assessment.

Tony Tyson's investment in gravitational waves began in 1971 when he built his own version of Weber bars. After years running bar experiments, the only event he ever detected was an underground nuclear weapons test in Alaska. A nearly 5 megaton nuclear weapon was dropped down a vertical shaft. On detonation the surrounding Alaskan surface rose a good 50 feet in less than a second and sent a tilt wave propagating around the Earth several times, ringing Tony's gadget in his Bell Labs facility. By the time LIGO was debated in Congress, Tony had moved on to other research areas, but he considered himself a supporter still.

When he received a request from the Subcommittee on Science to testify, he thought, "I better not get involved," until they threatened subpoena. Given less than a month before the hearing, he agreed to prepare an engineering calculation to assess the technological feasibility of the project. Tony spoke in support of LIGO, at the very least in defense of the elegant technological advances promised. To this day, Tony asserts, "If there is a new window on the universe, we should look." But he did summarize concerns over the scientific payoff, concerns that were shared by others less enthusiastic about pure technology and less comfortable with the risks. He compared the first generation of LIGO unfavorably to cheaper facilities with greater discovery potential, suggesting that astronomy would require future generations of observatories perhaps decades away and those generations were not covered in the escalating $211 million budget requested. He also complained that the big budget was for a facility with four users (presumably Kip Thorne, Rai Weiss, Ron Drever, and Robbie Vogt). This excerpt from his testimony made an unforgettable impact:

"Imagine this distance: travel around the world 100 billion times . . . a strong gravitational wave will briefly change that distance by less than the thickness of a human hair. We have perhaps less than a few tenths of a second to perform this measurement. And we don't know if this infinitesimal event will come next month, next year, or perhaps in thirty years."

Tony relayed to me that he regrets he did not notify either Kip or Robbie sooner of his compliance with the committee. He did begin to consider the implications, though by his own admission only in the final hour, and on the eve of the hearing managed to get them a copy of the testimony he would deliver.

"Actually," Kip Thorne says, "he sent the copy of his testimony on the eve of the hearing via FedEx to Robbie at Caltech, but it arrived after Robbie had left for Washington, so neither Robbie, nor any of the rest of us, were aware of what Tony Tyson would say until he said it to Congress. We were totally blindsided."

After Kip read his testimony, Tony received a difficult call from him late one night, and many sleepless nights followed. Tony summarizes, "Robbie had some interesting language. Kip was obviously very wounded. And therefore I was as well."

In an e-mail on March 16, 1991, Kip writes to Tyson in defense of his estimates of sources and speaks of "the caution with which I have tried to approach the issue." He goes on to say, "I strongly suspect, in fact, that your feelings, and those of other astronomers, 'that the gravitational wave strengths and rate of occurrence of hypothesized source have been grossly overestimated' [quoted from Tyson's congressional testimony, which also referenced an informal survey of astronomers Tyson regretted carrying out], are completely unrelated to the estimates that I have made in the LIGO Proposal and in the Astronomy and Physics Survey Subpanel Reports."

In a postscript to the letter, Kip adds, "I would be less than candid if I did not confess that I have felt personally deeply hurt by the 'gross overestimate' passage in your testimony. It has left me sleepless for the last few nights. I feel deeply that it is unfair. I have made enormous efforts over the last few years to be [impeccably] honest and accurate about the estimates. Please either help me to understand concretely where I have gone wrong, or help me to contain the damage to the LIGO Project and to my own reputation."

Three days later, Tony Tyson sent a fax to Robbie: "I HAVE CHANGED MY ORIGINAL WRITTEN TESTIMONY." All caps. He removed the word "grossly" and added the words "in the past" so that the amended testimony reads, "Most feel that the gravitational wave strengths and rates of occurrence of hypothesized sources have been overestimated in the past."

The addendum to Tyson's testimony closes, "On a personal note, I must say that this review has been quite painful; I have friends on both sides of this LIGO issue. Somehow we have to find the resources to support risk and innovation on all scales, from the cleverness of the individual researcher, to the larger initiatives for facilities guaranteed to produce wholesale science, to the risk and promise of occasional big science projects."

Robbie remembers, "Tony was really a shock because I didn't expect that. He is credible. He is a good scientist. We are on good terms now. Anyway, that testimony was devastating." The lobbyist leaned in to him to say, "They really butchered you."

Out of distrust for LIGO's big science character—usually the nature of ambitious physics accelerators, not astronomical "observatories"—there grew a sort of anti-LIGO movement, an opposition to the congressional allocation of funds. To set the scale, $200 million was twice the yearly NSF budget for astronomy. (Rich Isaacson counters, "This is a misleading comparison of apples and oranges—you are comparing LIGO construction, a multiyear facility construction effort, to an annual budget for research.") At risk was the health of smaller science with larger scientific payoffs. LIGO's argument, and the NSF's, was that the request would define a new budget line and therefore ensure in the long term more money for science. No dollars would be siphoned from the research that the NSF nurtured, and more would be available for development of visionary instrumentation in the future. Still, powerful astrophysicists from Princeton University, John Bahcall and Jerry Ostriker, opposed LIGO. "There was a conspiracy against me at Princeton. They were worried LIGO would take money away from astronomy. Noble reasons." Robbie shrugs.

Rai tells me that the word "observatory" in LIGO's name caused alarm for philosophical reasons (it's not an observatory until after you observe something) and for economic reasons (the competition for funds with other much cheaper observatories already discussed) and for sociological reasons (the project sounds more like physics than astronomy and hadn't the right to an astronomical title). Rai takes some blame for the name and wonders how things might have gone if they called it a "facility" or an "experiment." But LIGF and LIGE don't sound as good, you have to admit.

The negative campaign delayed construction. Robbie needed influential friends in Congress, and at first he had Senator George J. Mitchell (D-Maine), majority chairman, who wanted

LIGO in Maine. The LIGO team, with help from geologists at the Jet Propulsion Laboratory searched for two sites, and Maine was perfect but a bit more expensive than projected since more excavation and grading would be required. Mitchell promised to help raise the extra money with a special bond issue and a $6 million contribution from the state.

Robbie asked, "Why would you, a poor state, make a bond issue for something as abstruse as LIGO?" Mitchell's answer: For credibility. Maine wanted to lure other high-tech and bio-medical facilities and they figured nothing was as way out as LIGO. They wanted to use LIGO as a demonstration of their progressiveness and their dedication.

When Robbie made the presentation to Congress, his determined recommendation for the sites based on complex seismic and geological information were Hanford, in Washington State, and Maine. Robbie was set back when Walter Massey, then director of the NSF, refused to make the decision for the site selections on the spot and refused to discuss the crucial topic further. Sometime later, unexpectedly, Walter Massey urged Robbie to D.C. The selection of the sites was going to be announced in the Senate Building at a press conference, and he wanted Robbie there to back him up.

Robbie relays the exchange: "And I say, 'Walter, what did you pick?' He says, 'You'll find out when you get here.' "

When Vogt arrived in D.C. he learned the selected sites were Hanford in Washington and Livingston in Louisiana. Vogt protested, "Walter, you just pulled the rug out from under me. Mitchell will be furious." Mitchell *was* furious. Maine had invested heavily to support the site selection. Mitchell had fought hard for LIGO, but Maine was also scientifically the better site. Vogt eventually learned that the turn was specifically political. The Republican White House made the decision to punish Mitchell, the Democratic majority leader of the Senate. Although Vogt fought for Maine, he lost Mitchell as an ally in Congress, which still had not given him construction money. By congressional standards, the sum was in itself unimpressive, a bit of flux in terms of the country's overall budget. Less quantifiable but more important was the value in political currency.

Rich Isaacson of the NSF remembers the details differently. Many other sites were considered initially, from military bases to private land, from deserts to swamps, in geographies from Utah to California to the East Coast and plenty of locations in the middle. Robbie presented the NSF with an unprioritized list of more than one hundred possible pairings crossing nearly twenty potential sites. The director of the NSF, confronted by a list without rankings, assembled two separate committees to scrape the pairings with the sharp edge of several criteria, including relative orientation of the two detectors, seismic factors, cost, ease of procuring the land, and any other significant gauges that might be applied to reveal the winning combination. In the end, the director exercised his judgment and prerogative. Isaacson says without ambiguity, shaking his head, "The NSF makes decisions on the basis of science, not politics."

Regardless, Vogt needed a new ally in Washington. He turned to his lobbyist and said, "Get me an appointment with Johnston." She warned, "It will be difficult." And it was. Robbie is sure it cost her plenty to get that first twenty minutes. But then Robbie charmed those first twenty minutes into two hours. Senator J. Bennett Johnston of Louisiana became so interested in cosmology that he canceled his next few appointments and set in motion the future of an ifo in his state, the LIGO Livingston Observatory (LLO). Eventually cross-legged on the floor, Professor Vogt and Senator Johnston drew spacetime diagrams of the beginning of the universe and saw their legacies take on the gratifying subtlety of detail—deals made, sites secured, money appropriated. After two years of hard-fought political campaigning, Congress approved the allocation of $200 million to Caltech for the construction of LIGO.

Vogt says, "For that I take credit. I did get the money. Those were hard days. But by god, to win this kind of a fight . . . that was winning. I like to win."

Suddenly LIGO became the biggest project ever attempted at Caltech (this accounting does not include JPL, which can brag of megalithic missions like Voyager). Caltech scientists diligently focused on their own research, buried in their own labs, contentedly uninformed about academic politics first heard the acronym LIGO as a designation for a project that promised $200 million for a first generation of machines. Probably many disoriented experimenters emerged from their shops gobsmacked, hanging on that significant detail, despite the fact that Kip campaigned methodically to keep everyone at Caltech informed and supportive.

LIGO could begin in earnest. Ground could be broken (but wouldn't be for some time), buildings constructed, and in them over the next two decades the current instruments would come to life, assembled fine bone by fine bone, deconstructed and reanimated, hot blood-red light in its two gross arteries. But not before Robbie was fired.

Robbie knew that as big a number as $200 million sounded, it was bare-bones. He could finally dig into the project in his true style. No administration, just the best scientists in the world, and they would work seven days a week, sixteen hours a day.

It is conjectured by some on the faculty that with a paranoia-laced forcefulness, he galvanized new scientific staff and a fistful of postdoctoral research associates fresh out of graduate school. The team had an ungrudging leader with a vision. Spirits were elevated but charged. The scientific viability of the project was not unquestioned. There were resistance and suspicion from scientists outside the group, and maybe Robbie used the threat of that amorphous enemy to rally his modest horde, keep them close and closed and loyal, skunkworks style: a small, specialized crew funded and isolated in near secrecy to ensure innovation without restrictions. There is no expectation in a skunkworks that the team reports within any bureaucracy, and the organizational structure is missing the conventional hierarchies.

A reference to the Advanced Development Programs of the aerospace and defense corporation Lockheed Martin, the term "skunkworks" has some utopian tones, suggesting an unrestricted incubator. In Burbank in 1943, in around six months, Lockheed developed the first U.S. jet fighter, the P-80 Shooting Star, under a circus tent that accumulated unpleasant smells from a nearby plastics factory. The R&D technicians wisecracked that the odor matched imagined emanations from the moonshine factory Skonk Works in the comic strip Li'l Abner. The name stuck with modest permutation and became an alias for the Lockheed project.

Robbie's skunkworks management style was motivated by his archetypal hatred of authority. His contempt for executive supervision often drove him. He would accept an administrative job due to this loathing. Learning of the alternative to his nomination, he would think, "Not that idiot" and begrudgingly accept the position in order to spare everyone. He says, "Whenever I held office I was always convinced that the office above me was an idiot and that I had the most important job. And as I moved up the ladder I found out there was always someone above me that was an idiot." To avoid this recurrent problem,

Robbie was determined that there be no one above him, not even the NSF. They were to deliver the funds and stay out of the operation. There would be essentially no bureaucracy and no obligation to justify the rolling decisions the scientists made, to the NSF or to anyone else.

"If an authority wants power they have to persuade me I respect them. If they are bureaucrats, I will simply not respect them, and if I don't respect them I will not cooperate. I have not cooperated since and it has gotten me in a lot of trouble. But it has given me a personal comfort that that's what I wanted to live like."

Robbie explains, "Anyone who lived under the Nazis should hate authority."

Robbie describes his father affectionately as very sarcastic and very outspoken, a scholar, an Egyptologist, and also ardently opposed to the rise of the Nazis. His mother was nonpolitical. She was an industrialist, having inherited her father's business. Robbie stopped to butter me up a little: "Incidentally I have always had a prejudice in favor of women because I felt they were not getting equal rights, and not because I was a white knight or a good person. And the reason is my mother had been running a big industrial operation because she had been the only daughter of a man. . . . And I admired my mother. My mother was the most beautiful woman in the world and the most able woman in the world. . . . And she took me to her plants, which I understood. . . . And I had female professors who were very good so I was prejudiced in favor of smart women."

"It's a good prejudice to have," I congratulate him.

The Caltech archivists already have the keys to his office and are collecting official documents. There were industrial-sized garbage cans in the room to cull material—for disposal or transportation, I'm not sure which. The evidence of Robbie's existence will be shelved in the archives for scholars to peruse. Robbie became a father, an accomplished scientist, an influential scientific leader. Moody and intimidating. Fierce but fragile. Antiauthoritarian with an emphatic demand for loyalty. On May Day 1945 he began to rewrite his life's story, but his history still impels him.

Many Germans suffered under the Nazis, he wants me to understand. "Other Germans were not victims like the Jews, but life under the Nazis was indescribable," although he tries. "At that time the Nazis had a very powerful method of discouraging people from opposing them. When they arrested a man they arrested his wife also." The orphaned children of political prisoners were placed in cadet schools, then sent to fight. A fourteen-year-old could be charged with an entire troop of child soldiers. A child had "a body that could stop a bullet and that's all they needed." The children would have no military training. They had pickaxes and shovels. When the invasion happened in 1944 and the British broke through, the German Army issued bazookas and rifles and would send these "kids" into combat. None of them would survive. Robbie's rage and deep-rooted contempt for authority surfaces vividly as he describes the atrocities. "The government had power over its citizens and used it ruthlessly."

This digression was unsolicited, almost sudden. Just as impulsively he seems to want to abate the momentum. He flips over what I take to be an imaginary hourglass and then slaps the table heavily. "On May eighth, 1945, the world turned over for me. I started fresh."

Now in his mid-eighties, he looks at me through wilted orchids. He tells stories that swing from funny to terrible. He is forceful and fragile both ("Mornings are always hard," he mentions) because obviously the damage of those years that he wants to erase is indelible. He wants to remember his contributions to society as a scientist: Voyager, the Keck Observatory—the largest optical and infrared telescopes on Earth. (Robbie was central to securing the money for Keck.) Robbie wants to remember LIGO even. He feels a continued obligation still as a protector of science, of his adopted country, of its citizens and its ideals.

The beginning of the end of the fifteen years he wants to forget came with the Allied invasion. I ask my first questions about his childhood and the recollections noticeably pain him. He glances around but the search is internal. He wants to move away from this. I am confused about the facts and my queries are simple. He tells me some things, brief and laconic. The details are clipped, the delivery cautious—he apparently decided on a short bridge of minimal facts as the simplest, most direct way out of this terrible maze. "I will not be a celebrity or made a hero by anybody. I want to be anonymous."

He looks at me for a few breaths, the only pause in the more than five-hour conversation. I don't know, but I feel he is assessing my reliability as a confidante. He looks at me with more focus than at any other time during our meeting, maybe searching for clues to determine if he should entrust me with a reply. I return his exploring search, wide-eyed and expectant. Quietly—and Robbie doesn't say anything quietly—he says, "You asked me if I aspired to office. No. I detest authority. And the exercise of authority . . . it corrupts you."

12

Gambling

Stephen Hawking makes notoriously bad scientific wagers. He's won none of the many public bets he's made. He bet against Caltech theorist John Preskill that information never escapes a black hole, not even in the eponymous radiation that Hawking himself discovered. Then he conceded the bet, although many—probably including the winner Preskill—would say prematurely. Kip was in on the bet, on the same side as Hawking, and has not conceded.

Hawking bet that the Higgs particle, which glues together the pieces in the puzzle of our material reality, would not be discovered. The experimental particle physicist Leon Lederman famously referred to the Higgs as the "Goddamn Particle," a moniker his publisher resisted so that his book was titled *The God Particle.* Unfortunately the flourish has stuck. The Higgs was found, Nobel prizes were awarded, and the discovery was both a disappointment (is there nothing more?) and a triumph (they did it!). Hawking paid the hundred dollars to his colleague Gordon Kane.

Hawking made his oddest wager about killer aliens or robots or something, which will not likely ever be resolved, so that might turn out to be his best bet yet.

To critique Hawking as a notoriously bad gambler—not bad as in compulsive, just bad as in profitless—might be to miss the point made obvious in his most famous wager. Hawking bet Kip that there was no black hole in Cygnus X-1, the brightest X-ray source seen consistently from the Earth (though not intrinsically the brightest known X-ray source). The bet was made in 1974, ten years after X-rays from Cygnus were first detected. Hawking was already deeply invested in black holes, having secured his fame with his realization that black holes could evaporate. Sometimes Stephen hedged his bets just for fun. The phrasing of the wager began, "Whereas Stephen Hawking has such a large investment in General Relativity and Black Holes and desires an insurance policy, and whereas Kip Thorne likes to live dangerously without an insurance policy . . ." In 1990, Stephen and his entourage stormed Kip's empty office to concede—Kip was in the Soviet Union at the time of the breaking and entering. The promissory note was sealed with Stephen's thumbprint. When he did pay up in the form of the agreed-upon subscription to a sexy magazine it was "to the outrage of Kip's liberated wife." At least that's the story propagated since. "I was never outraged," says Carolee Joyce Winstein, Kip's liberated wife. "My reaction was one of surprise more than anything . . . because I thought the women's movement was well under way in sensitizing folks about these things. Clearly I was mistaken. This was likely too heavy for the press and so it got reduced to the stereotypic 'wife was outraged' story." Carolee, who is not at all prudish, finds the whole thing rather amusing.

Kip is a more successful gambler than his friend. He claims

to win all his bets as long as they don't have a date on them. One dated bet he lost to the prolific astrophysicist Jerry Ostriker, who, incidentally, was a contributor to the theory of X-ray emission from Cygnus X-1.

As Robbie Vogt fought to secure funding in Congress, Kip Thorne campaigned on the scientific front. Jerry Ostriker listened to an enthusiastic talk Kip gave to a Princeton audience in the 1980s. He didn't want to give Kip grief during his lecture, but he was thinking, "Where is he getting these numbers?" The numbers referred to anticipated sources of gravitational waves loud enough for LIGO to detect. Ostriker was sold that gravitational waves would be generated by astrophysical systems. He was not sold that they were loud enough or plentiful enough to justify Kip's enthusiasm.

Kip has heard this before, the accusation of optimism, which he patiently refutes with references, documents, and published graphs. "There is a 'cherished beliefs' line"—he points me to a figure in an article that he published in 1980—"that answers the question, 'How strong could the waves be without violating our cherished beliefs about the nature of gravity or the astrophysical structure of our universe?' It corresponds to a very loud sky indeed! I never claimed that would be the *actual* wave strengths." Kip goes on to say in that 1980 article, "However, currently or recently fashionable models for the Universe predict that the strongest [burst signal] should lie far below the 'cherished beliefs' line." And while some of those fashionable models have since been reclassified as unfashionable, there was consistency in the push for detectors capable of delicate sensitivities, in the zone of the sensitivities of today's advanced detectors. (There were even T-shirts made for a 1978 conference that read, "10^{-21} or bust.")

Jerry Ostriker and John Bahcall, also from Princeton, were

probably the most strident LIGO critics. Kip, the charming, cogent LIGO campaigner (they might say propagandist), had to convert his colleagues and Congress, which would not give the thumbs-up to a big and long-term science mission without some surety, some certainties. Kip could make a solid scientific case for reasonable astrophysical signals that LIGO could detect with high probability. They will hear *something*. Probably. Nearly guaranteed. But even now Kip won't say absolutely guaranteed.

Whatever the earlier impressions, the realism of some sources is no longer disputed. The compact binaries are those certs for LIGO, the surefire sources for the Earth-based gravity observatories, insofar as we know that they exist even while plenitude remains uncertain. The "compact" descriptor refers to collapsed dead stars: white dwarfs, neutron stars, and black holes. They are compact—a lot of mass in a very small volume. And they are dead—they no longer shine terribly brightly, if at all.

When Rai first imagined LIGO, the surety of sources was not so sure. Twenty years after Wheeler conceded and lent black holes a name, highly respected astrophysicists were unfazed by theoretical evidence. They, rightly, expected better: empirical evidence. But even as that evidence accumulated, there could always be another explanation for the observations than a central black hole. Alternative explanations became increasingly more elaborate (maybe a gas cloud is arranged just so to distort the data, and other contortions). "The opposite of Occam's razor," Rai says. "Explanations so complicated and arbitrary . . . I mean"—a dismissive swat says the rest. "I can't go to MIT asking for money to detect black holes when the most esteemed members of the faculty don't think they're real."

Still, there was an excitement spreading through the ranks.

The discovery of pulsars convinced many scientists that neutron stars exist. The discovery of the Crab pulsar in a bright supernova remnant swayed the community toward the opinion that neutron stars were the end state of gravitational collapse for at least some stars. There was evidence for black holes from bright X-ray sources like Cygnus X-1. And the clincher: the Hulse-Taylor pulsar indirectly exhibits loss of energy to gravitational waves. With so many convinced that stars died as compact objects, there grew a surety that there would be sources. The question became, How many?

White dwarfs and neutron stars are extremely faint. We cannot see them if they're far away, extragalactic. We can see evidence for them in our own Milky Way galaxy, which is about 100,000 light-years across. The nearest big galaxy, Andromeda, is about 2.5 million light-years away. We can see supernovae in distant galaxies but, if they are millions or billions of light-years away, not the pale collapsed core they leave behind. We have every reason to extrapolate from our knowledge of our own galaxy to other galaxies. There are vast populations of stars in the vast populations of galaxies in the observable universe. There must be dead remnants among the hundreds of billions of stars in each of the hundreds of billions of galaxies. But extragalactic compact objects are simply too faint to detect with telescopes.

With such vast populations of compact objects, LIGO aspires to detect them with a gravitational-wave observatory even if they can't be seen with telescopes. A compact object just sitting there by itself will not emit gravitational waves. A prone drumstick doesn't bang the drum. The mallets have to move. The concentrated masses have to accelerate to impart energy to gravitational waves. The Hulse-Taylor pulsar accelerates in its orbit around the other neutron star. Very possibly most star systems are born in pairs and die in pairs, although the supernova explosion that forms the collapsed core can sometimes eject the companion. An entire class of neutron stars and black holes that are sought by LIGO are those that end their lives in a pair. (White dwarfs in binaries ring spacetime at notes below LIGO's range.) The compact objects orbit each other, and the accelerations of those mallets drag and swirl the curves in spacetime around them, emanating gravitational waves.

Here's where the pitch to Congress gains some momentum. Compact binaries generate ripples in spacetime in the wake of their orbit at the expense of the orbital energy, so they spiral a bit closer together—they inspiral. With each orbit, the dead stars get a little closer and the orbits take a little less time.

All astronomical binaries, not just compact ones, will emit gravitational waves. Barring other orbital changes due to solar system effects, the Earth will slowly spiral into the Sun with loss of orbital energy to gravitational waves. The Moon will spiral into us and the Sun into the center of the galaxy, but all of this is absurdly slow and the gravitational waves are imperceptibly faint. It would take way way way longer than the age of the universe, for example. The Sun will die first. The Milky Way will collide with Andromeda first. The chances that our species will be here operating gravitational-wave interferometers to detect the apocalypse shouldn't get very good odds. Hawking might make such a bet for sport.

But LIGO will be able to hear the end stages of inspiral of a pair of compact objects. Imagine the final moments of a black hole collision. Two holes in space, each maybe 60 kilometers across, executing hundreds of orbits every second, moving at a significant fraction of the speed of light before crashing and merging. Their motions will ring space loud enough for us to hear the calamity when the modulations roll over the Earth.

Only in the final throes will the gravitational waves be loud enough by the time they hit the detectors. The chances of catching a compact binary with lifetimes in the billions of years in its last fifteen minutes is dispiritingly unlikely if observations are limited to our own galaxy.

In the Milky Way, there may be one neutron star collision with another neutron star every ten thousand years, although these predictions are still very uncertain. There may be one neutron star collision with a black hole every few hundred thousand years. There may be one black hole collision with another black hole every couple million years. And so it would be very silly to spend fifty years building LIGO to record compact binary collisions in our own galaxy.

LIGO must record the ringing of space originating from within millions of galaxies in order to record black hole collisions on a scientifically reasonable scale (say within a year of commissioning). But other galaxies are far away, so LIGO needs to probe to great distances to scoop many more candidates into the observable reach. But the farther the fainter, so even though the initial LIGO generation was operational for six science runs (times when the machines were fully operational and recording data), the machines could only hope to detect neutron star binaries within roughly 45 million light-years, reaching out to the nearby Virgo cluster of galaxies and black hole pairs a bit farther. Seems far, but not far enough. None were heard.

After Kip's talk at Princeton those decades ago, Jerry asked Kip the question on his mind: "Where are you getting your numbers?" The odds that a first generation of detectors would hear something were only good if sources were plentiful. Otherwise, odds were bad. Theoretical assessments of the abundance of sources were wildly uncertain. Kip would say that the highest numbers were at least respectful of physical law, but the lowest numbers were always their targets. Jerry has a pedigree in astronomy and the cherished belief numbers were distracting and violated astronomical realism.

Kip wagered nearly thirty years ago that LIGO would detect gravitational waves by the turn of the last century. Jerry wagered quite confidently that they would not. There were sensible scientific qualifiers that Jerry appended, like at least two groups had to agree that gravitational waves were detected and that each group had to agree the other performed the analysis of the detection correctly. But the qualifiers turned out to be rather unnecessary. January 1, 2000, came and went and the initial generation of LIGO, while just completed, was not yet taking data. Jerry claims, "My bet disappeared from his wall at some point."

"For the record," says Kip, "I think the bet hung on the wall continuously except for a few days when I took it down to sign my concession."

Jerry says he didn't demand payment immediately. Instead he persistently inquired among Kip's friends, "How's Kip?" And in this patient and dogged manner let them carry the reminder to their friend that the bet was owed.

Kip counters, "Jerry obviously misremembers. I lost the bet on January first, 2000. I have a nice note from Jerry, handwritten and dated 4/18/00, which reads, 'Many thanks for the gracious note and the VERY fine wine! I, Jim Gunn, Bohdan Paczynski, Scott Tremaine, and Martin Rees have all drunk it and toasted to your health and the success of gravitational wave detection in general and LIGO in particular. With all best wishes, Sincerely yours, Jerry O.' "

Jerry Ostriker is just one of many astrophysicists who were "quite cross" about LIGO. Jerry specifically refers to procedural irregularities. The important Astronomy and Astrophysics Decadal Surveys prioritize missions and set the program for the coming decade. A significant dimension of scientific self-governance, Jerry has served on three of them. Never was LIGO on the list, although there is some discussion whether John Bahcall refused to even consider LIGO. Jerry was cross that all of the major projects over the past many decades were vetted by the Decadal Survey, but not LIGO. He was upset, as were others, that money was spent on tunnels and not on graduate students.

On this issue Kip protests, "A key point is that LIGO was being funded out of the Physics Division of NSF, not the Astronomy Division. The Astronomy Division was never in the loop in any significant way; it was always Physics.

"LIGO was vetted by the Physics Decadal Survey Committee. . . . It was from NSF Physics that LIGO was funded until the new funding line was created," Kip continues. "A major aspect of getting LIGO approved was the huge number of reviews that we went through, with review committees populated by hardnose physicists such as Garwin, from the mid-1980s onward. There is no way that NSF would approve LIGO without such reviews."

One critic who requested anonymity suggested that Kip's Mormon upbringing groomed proselytizers. Although Kip outgrew the tedious moralizing, the sexism, and the religiosity of his Mormon roots, the suggestion was that he never shed the urge to proselytize for a righteous cause and LIGO was that cause. Why was LIGO funded despite the staggering cost and the tremendous risk? Because Kip was a very charming and very convincing advocate. He was also very thorough scientifically, clear in his analysis and reviews of the state of the art, and respected for his integrity. Kip could make you believe.

The initial LIGO generation was a technological success.

But still there was no detection. To record the faintest chirps from billions of light-years' distance requires ratcheting up the technological challenge. Advanced LIGO is designed to extend over a billion light-years, and out to that distance we can reach millions of galaxies. With careful estimates of populations of stars, their sizes and life spans, astronomers try to predict the number of compact binaries that will merge within hearing distance. And so we have the guaranteed LIGO sources. They are guaranteed to exist, although even current estimates of the number are disputed, with pessimists and optimists equally vocal. There remains no guarantee that a pair of compact objects will collide within a detectible range in our lifetimes.

We rely on Nature's generosity to provide sources in great-enough abundance that the machines will hear the soundtrack of the universe within reasonable scientific runs—a year or two, not twenty or thirty. There very likely won't be the will to maintain operation of the facilities if a detection is not secured within a few years. LIGO also needs to do more to justify the investment versus rewards calculus. LIGO has to "do astron-omy." And that provocation has many worried and many in a bit of a fever of neck-aching calculation to expand the range of astrophysics that LIGO can explore. LIGO still has many detractors and will need to justify its existence to the entire astrophysical community. That remains problematic for the coalition. Will the return be enough? Will there be enough sci-ence to merit the expense?

These days, LIGO is the only machine approaching detec-tion capability, and so Jerry Ostriker's condition—that at least two groups agree that gravitational waves are detected and that each group agrees the other has performed the analysis of the detection correctly—cannot be met. However, Ostriker says he'd be convinced if there was a coincident detection with a telescope. An event that is both loud and bright—a supernova or neutron stars that burst brightly as the condensed superconducting magnets crash. Now that Advanced LIGO is essentially here, Jerry is interested and open, and like many scientists he will enjoy the scientific rewards without acerbity. Still, he's placed another bet with a LIGO scientist, though not Kip. Jerry bet that gravitational waves, confirmed by a corresponding observation from a telescope, would not be detected by January 1, 2019.

In the far future, which promises to be vastly longer than our past (like a googolplex of years to our future versus 13.8 billion years to our past), all of the stars in the universe will have run out of fuel. Those that can will collapse to black holes; eventually everything will fall into stellar-mass black holes, and those black holes will fall into supermassive black holes, and then all of the black holes in the universe will eventually vaporize into Hawking radiation. This will take a very long time. ("Eternity is a very long time, especially towards the end.") All of the Hawking radiation will dissipate in an ever-expanding cosmos, unable to fill the swelling void, and the light in the universe will go out. Eventually every particle will find itself alone, no bright sky above, no luminous solar systems below. For now, we're here and the skies are bright, if somewhat quiet. The gamble is that the skies aren't silent.

13

Rashomon

By 1987, Ron Drever had accepted the dissolution of the Troika and its succession by a hierarchical structure with Robbie Vogt as director. He simply had no other choice. Naturally, before the consummation, he did fly to D.C. for a handful of meetings, plaintive with Rich Isaacson of the NSF, who patiently but persistently instructed Drever ("point-blank," says Kip) to accept a single director with the authority to make a decision or the project ended there.

In the beginning of Vogt's term as director, Drever says he was not entirely unhappy, although Vogt did make many organizational changes and, unconcealed, did cart around a difficult temperament. The operation overall became more professional, more effective, and substantially funded. At first Drever thought Vogt was "quite okay," although he was sore that not all of his technical decisions were implemented immediately. But Ron had the impression that Vogt's was to be the only voice the NSF would ever hear directly from the project. (Others regard this as an exaggeration, citing group presentations to the NSF.) Ron slowly felt his control erode. He complained that it was hard for anyone else on the team to fully understand the status of operations at all levels, Vogt so tightly controlled the flow of information—Ron's impression, not, for instance, Kip's. He worried that the scientific progress was also slowed in order to write the proposal. Kip had to step in and write substantially, since Ron hated writing, which also led to some friction. Ron criticized: Some of the statements were too optimistic, too convincing, particularly around estimates of instrumentation performance. He suspected that Kip might have been annoyed with him, thought him fussy. Ron wondered in retrospect if the tension must have grown in the background without his realizing.

In 1991 Stan Whitcomb was deputy director, aide, and ambassador for Robbie Vogt—and part therapist. Stan was one of the original experimenters on the LIGO design in 1980 but left in 1985 when recruited by headhunters in the aerospace industry, given a backdrop of uncertain prospects for his own academic career, tethered as it was to the then highly uncertain future of the project. (Others attribute his departure to tensions with Ron.) Stan was lured back, he says, by the memory of the most fun era: just tinkering in the lab, innovating. In the earliest days, the early 1980s, there were only three or so people in the laboratory allocated to R&D. Everyone knew what to do and got to do everything by hand. He returned by 1991 as deputy director to a more optimistic era, but also more serious, less exploratory. Stan describes Robbie as a harsh but effective director. "Where Rai, fiercely analytical, and Ron, fantastically intuitive," were in a stalemate under the Troika, Robbie forced everyone to work together effectively. But still Ron Drever and Robbie Vogt couldn't get along.

All of the elements were in place finally to free LIGO from

the gate, the team no doubt fidgeting with impatience after three years of an industry study, the year-plus commitment to the coherent proposal, the subsequent two-year battle with Congress. The next engagement should have been a confrontation between them and nature, humans against the elements. Confidence in astrophysical sources grew, the bets decided in favor of black holes and neutron stars. The group must have been agitated, ready to break out in big long strides. Instead, the external skirmishes fought and won, Robbie ran a finger along the internal fissures, previously too thin to draw attention away from the outward crises. As with a tongue along a cracked tooth, the finest fracture might have felt like a chasm.

Despite shared key points, the main characters construct contradictory renditions of what came next. Many people declined to discuss the ensuing conflict. Not least among the reservations listed was an unwillingness to publicly criticize Ron, who may have seemed inadequately equipped to manage disparagement even under optimal circumstances. In many hours of tape-recorded interviews over five sessions in 1997, conducted by Shirley Cohen for a Caltech oral history, Ron recounts, unhesitating, his own perspective of events. In my interviews with principal LIGO scientists, those who were willing, if reluctant, to comment relay accounts fairly consistent with one another but different from Ron's. And nearly everyone requested anonymity.

From the 1997 recordings emerges the following version conveyed here in deference to Ron's viewpoint. He describes the troubles that began in the late 1980s under Robbie Vogt's leadership. Maybe not with the full force of his rancor, but with enough of it, Vogt began to attack Ron in the weekly group meetings. "Particularly he kept accusing me of not using the scientific method. And this hurt me tremendously." Ron attributes the techniques he learned in the United Kingdom to Rutherford's influence. He cut corners without compromising performance, did many experiments very quickly, skipped extraneous details, and moved very fast. His practices should not be misinterpreted as slipshod. And there is no doubt that Ron invented an impressive collection of ingenious experimental techniques and designed significant original elements still crucial to the machines today. He protests, defensive, that his methods had put them ahead. He could work twice as fast as traditional groups and, always frugal, for less money. Other scientists often seemed conventional to him, unable to make the kind of jumps he could make after thinking very hard over a problem. Vogt just did not understand the mechanics of Ron Drever's genius.

"I would take a step that wasn't obvious, and it would work. Robbie would say, 'He guessed!' Well, I didn't guess. My intuition was very powerful—it *is* very powerful . . . but I found it difficult to explain it." Ron continues, "But he became more and more against me, and I didn't know at the time quite why."

The attacks at the weekly meetings were worrisome and difficult and, for Drever, baffling. Unsure how to manage the assault, he would often stay quiet. Then Vogt removed Drever from his role in charge of the lab. "I was shocked—I remember, I almost broke down."

Peter Goldreich, professor emeritus at Caltech and Princeton, recounts, "I remember Ron telling me one time, 'This is awful. This is awful.' Robbie would be yelling at him. And I'd say, 'Why don't you just walk away if he starts yelling at you?' Ron said, 'Could I do that?' And I said, 'Of course you can do that. You are a professor here.' . . . I couldn't believe that Ron was so naïve."

Peter was part of the faculty that originally supported Ron

Drever's hire to run LIGO's experimental program. "It was clear to me that he was a funny guy from the few times I met him . . . he was totally unworldly and totally involved with physics and very intuitive. . . . I knew from previous experience that Robbie was capable of developing irrational hatreds for people and was very effective in convincing other people that the objects of his scorn really deserved it. That's what really made me feel bad . . . what happened to Ron, I feel responsible for."

On the third of five interviews Shirley Cohen conducted with Ron from January to June 1997, he wanders toward the culmination of the story, sometimes on looping circles for a repeat transit of earlier topics. She seems exhausted and encourages him to wrap up. "We're getting to the point where I got flung out!" He laughs and the second side of the tape snaps to an end. Getting to the sacking would require four more sides of tape.

"I'm slightly dyslexic, or something like that," Ron admits. Since he felt at a disadvantage assimilating information, he wanted to tape meetings to review later, but Vogt wouldn't allow the recordings. In the oddest meeting of all, according to Ron's memory of the incidents, Vogt gave him two rules. "The first and strangest of the rules was that Robbie Vogt and I must never be in the same room at the same time. That's what he said." The two witnesses recorded written versions that were less crazy. If Ron entered the weekly meeting, Vogt would walk out and the meeting would be canceled and thereby Drever would destroy the work of the project. "The other rule was something like the following: That I shouldn't make any use of the project facilities—Xeroxes, telephones, or something like this." Years later, as he relays the story, he sounds freshly baffled, totally bewildered, bemused, with the distance of time maybe even amused. "That was a kind of significant meeting,

I think, because it really was so weird. That was just before, I think, this conference in Argentina."

Ron was forbidden from presenting his work at other venues, universities, or conferences. Out of concern for the project, miserable and anxious, he complied for the most part. "I didn't know what was normal in this country. And I was more and more learning it was very different from what I was used to. None of this could have happened where I was from. But okay, I didn't know what was the norm."

The fateful exception to his compliance was in connection to a conference in Argentina. In 1992, Drever intended to present his work with Brian Meers, a colleague from Glasgow, who had done analyses of Drever's ideas, developing the theory for recycling laser light in the ifo. (Colleagues on the ground remember his collaboration with Meers differently than Ron does. They recount his resistance to the young Meers's ideas and Ron's frustration with the emphasis those ideas were given.) They were preparing a joint paper when Meers was killed in a climbing accident. Practicing for a holiday in the Alps, Meers and a colleague, Patrick Grey, roped together on the highest mountain in Scotland (Ben Nevis) during unsuitable weather, fell from a precipice. "That was a terrible blow to everybody, of course, and to me, too. We all liked him, and he was just killed." Ron felt some urgency, he suggests impelled by the tragedy, to present this work at the 1992 conference in Argentina. Forbidden by Vogt to do so, he presented anyway.

The day he walked back onto campus, Ron was tossed off the project.

The Caltech administration, not specifically Robbie, fired Ron in the end. On July 6, 1992, Robbie Vogt followed with a memo to the LIGO community and a fair fraction of the

Caltech community. Ron Drever was no longer part of LIGO. He was not permitted to extract his personal effects from his LIGO office unless accompanied by a member of the LIGO staff.

The grievance had become so amorphous, so without edges, that it bloated into the past as much as the future. Ron wandered through his memory even of good times, the first five years, when he thought the work was going quite well, to discover in those recollections previously undetected hints of sabotage. Bolstered by rumor, he learned that Vogt had complained to the administration about him during that half decade, as though preemptively—Ron believed with the intent to destroy his reputation.

Maybe Vogt hoped Drever would just resign, go back to Glasgow, as any person with a more conventional and reactive psychology might. Ron's unconventional mentality could have made him oddly resistant to ordinary manipulation. Push him unremittingly in one direction and he could still move unpredictably in another. Apparently his primary reaction to the pressure was an impassive bewilderment instead of a spiteful resignation. He lived for the work, for the lab, for the implementation of his inspired ideas. LIGO was centered at Caltech. There was nothing else for him and nowhere else to go.

The door connecting his office to his secretary's was sealed up one day. Not locked, but walled up. Builders were brought in and left only the vague impression of the symbolic door in what Ron criticized as a shoddy job. His secretary was removed (to where? a basement?), and all Ron could think was, "This is terrible."

"There was a short spell when I was told not to come into work." He says this last part so slowly, with such sad disbelief. There are rumors that Peter Goldreich climbed into a window to let Ron into his office after the locks had been changed. But that's just a rumor that Peter wouldn't substantiate. Peter says, "And then . . . I got a memo that Robbie had written to the LIGO community. . . . This really annoyed me." Peter waves me away when I ask if he really climbed in the window, not as if to say this is ancient history, but because he's still pissed about it. "I think what Robbie resented more than anything else was that he, Robbie, this great engine of progress, was managing the project and this sad-sack little round man, Ron, was going to get the credit—and maybe even a Nobel Prize. And of course that nobody would appreciate Robbie—he had a problem with things like that. . . . I didn't really understand that, because we all admired Robbie. So I told Robbie at the end that he ought to resign, because he was going to be fired eventually. He would never win. He couldn't survive this type of thing. In the end, he would be fired."

Vogt says the real crisis broke when Drever "started calling up everybody saying I was mentally deranged and I was building an instrument that could never work." (Drever expressly denies this accusation.)

Rai Weiss was willing to go on record. "Ron was made persona non grata in the project. He was not allowed to come to a meeting. And it was way beyond what was needed. And the Caltech faculty was totally alarmed by this and they said you can't do LIGO without Drever. Here was a faculty, much of which felt that Ron was a great genius—I mean, even Kip felt that way—and that Robbie was not behaving in a way that was protecting that genius. The genius was complaining."

"So I just couldn't understand any of this. It was so strange," Ron says. Eventually with the encouragement of his few allies, Ron filed a grievance that was taken up by the Academic Freedom and Tenure Committee, importantly a committee independent of the Caltech administration. Of the committee report, Ron says, "This was basically supporting me very strongly. . . . It's a good report. It says really tough things . . . my academic freedom had been violated." Still, nothing happened for years. Ron was unwelcome on the premises. He was "frightened" to go into the LIGO buildings.

Unfathomably, he sought reinstatement in LIGO. During one meeting of an oversight committee involving outsiders, Drever expected to present a defense of a scientific proposal and Vogt to present a counterproposal and that a resolution between them would be determined on the basis of scientific merit. But instead, Ron was dismayed to enter a room of Vogt's supporters who stood up one after the other to deliver "personal attacks." Ron says, his voice deepened with emotion, "These people had been my friends."

There are other sides to the story, naturally. From those willing to comment on the condition of anonymity emerged a somewhat coherent rendition that leaves Ron's account a bit apart from the rest. The quotes, given here without attribution, compile the opposing view. "Before Robbie became director, Ron had already alienated most of the members of the LIGO team, and in subsequent years the alienation grew. . . . Among the factors contributing to the team's alienation from him were his effort to maintain full and sole control of his group's research, using other scientists as his assistants and rarely granting any significant responsibility or authority to them." "Ron was adamantly wedded to what he himself called his own 'nonstandard strategies of research' . . . based on his own intuition and no analytical analysis, and he would not or could not lead a systematic approach. . . . In 1988 and early 1989, Robbie tried to impose more standard systematic methods. Ron tried to block this and tried to prevent team members from following the systematic methods." "Ron was highly disorganized and had great difficulty making decisions, bringing issues to closure, and meeting deadlines—shortcomings that severely impeded leadership of a systematic, many-person research effort." "As Robbie—forcefully—tried to take control of the project . . . Ron fought him—not head-on but going behind his back in a variety of ways and 'gumming up the works.' This led Robbie to behave in some of the unfortunate ways that Ron describes." "The sealing of that door was part of a set of modifications to the secretary's office that Robbie had requested be made . . . the modifications, including sealing off the door had been discussed with Ron before the modifications began. Evidently, Ron did not remember that discussion." "The rekeying of Drever's office was requested by Robbie so that he (Robbie) could not be accused by Ron of going into it and taking things . . . this rekeying was discussed with Ron . . . he could get a new key. When Ron arrived one day and found himself locked out of the office, he evidently forgot the conversation." "From among twenty-five academic freedom violations alleged by Ron, the Academic Freedom and Tenure Committee concluded that there was one actual violation and two incidents of infringing his rights." "I think I can say this without being libelous: No one could deal with Ron."

The entire protracted episode became worthy of a name: "the Drever Affair." Rai says, "Both Ron and Robbie would demand loyalty of the deepest sort. 'Loyalty' was the right word. Are you with me or are you against me? Ron was questioning Robbie's technical decisions, and Robbie felt that he was more than just an administrator. I mean, I think if you picked a thing that he was sore about—if you ever said to him that he was a manager, he would jump all over you. [He'd say,] 'I'm a physicist, and I can think about things like this, like anybody else can.' I respected that, because he's not dumb. And Ron wouldn't give him that, okay? And that's, I think, fundamentally what finally happened to them. Ron pushed a button that made Robbie feel like a second-rate person, and Robbie couldn't deal with that."

Rai continues, "All of a sudden, we found this terrible polarization here. In the meantime, the real sin was going on. The real sin was—and I was still trying like hell to protect Robbie— but the real sin was that the project was not progressing."

The final conflict came in 1994. "It happened after we had started the contracts with Chicago Bridge and Iron, who built the tubes. I was the scientific advisor for that. Robbie had a fit—a public fit—with the NSF guy who was there to see the beginning of it. And it was embarrassing for the project.

"The NSF guy asks a question that to Robbie sounds hostile. To me it sounds like a perfectly sensible question. And Robbie threw a fit. I never saw a fit like that. He threw a temper tantrum. He turned red. Robbie's fairly big, shouting at the little NSF guy. And Robbie said, 'You cannot do this to us. You have to shut up.'

"The president of CB&I and all the engineers, they looked at each other and tried to figure out, 'Who is this madman jumping all over the NSF guy? I mean, that's the guy who's got the money. What the hell is he doing?' And I remember that that's when I broke with Robbie, and that was very hard for me. I'll tell you, it's probably the hardest thing I've ever done. I really feel I hurt the man.

"I said, 'You're in trouble all around, and I cannot protect you anymore. . . . It's time that you left. You've served your function. I hate to say it.' And Robbie then got into one of these terrible depressed funks where he looks like he's about to die. He turned into an absolute . . . like a skeleton. His whole visage changed. He turned white. We were in this car together. Nobody said a word.

"When we got out of the car I said, 'I'm sorry, Robbie.'

"And he said to me as I walked out and we were about to split—I was going to my plane and he was going to his plane—'You always see things wrong.'"

Stan recounts, "It coincided with the Northridge earthquake. Just a coincidence," he assures me, "but I remember we were in D.C. to get our asses chewed out by the NSF, to beg for mercy, and the news of the Northridge earthquake was on TV that morning."

Rai was also in D.C. at a meeting around that time: "Robbie was interrogated by the NSF. It was an absolutely gruesome scene. He tried to defend his decisions. He shouldn't have tried to do it. Why he got rid of Drever. Why he sat on the money. Why he didn't appoint a larger project office. All they did was read him the committee reports. And Robbie sat there like a dead man. And that was the end."

In Robbie Vogt's defense, Kip elaborates on the major things accomplished under his leadership. Vogt systematized the R&D so that an effective team gained ground in the design and implementation of ifo elements. Essentially, he made a proper LIGO research program. He oversaw the selection of the sites and the design of the vacuum system and beam tubes. He forced difficult decisions on the optical geometry and the laser. He facilitated a first cut at the detailed design of the initial interferometer. Vogt also secured approval for LIGO in all respects, from reviews to congressional concept. (The final construction funds had not yet been released.) Vogt made them a single team in pursuit of an attainable goal.

Robbie shrugs now. "I was a gambler. I was convinced I could build it myself." Aware that his reputation preceded him, in the fifth hour of our conversation Robbie says, part defense, part confession, "The mistakes I made were because of the information I had at the time." This first part was emphatic. Then he adds, smiling, "And because of my temperament."

Ron Drever was ousted from the core of the project. He was given roughly a million dollars and space from Caltech for his own research and a new laboratory that would be problematic from the start (lesser facilities, disruptive location, unsuited for renovation). In 1997, Drever's lab still was not properly finished or even funded. He watched in frustration as construction of LIGO began at the sites in Washington and in Louisiana. He said about the small experiments he was able to perform that he had "very much a feeling that this is only second best, not as important as the real detection of gravitational waves. And I feel very much handicapped—almost being forced, for reasons I don't understand, not to make my full contribution to that. . . . I can't help feeling . . . that I could have done more, that I could be doing more."

Rai says, "That whole episode is the bad part of LIGO. Ron Drever is a tragedy. Neither Robbie nor Ron ever really recovered. Nobody wants to resurrect this stuff. It's unfortunately in the public record now. But it doesn't have to be in your book."

14

LLO

My lord, they sure are sweeter here in the South. I don't care if it's deep down goodness or reflex. Jamie's voice has a slight memory of the South, having grown up in Atlanta. He collects me from the New Orleans airport on his way out to LIGO Livingston, closer to Baton Rouge. We drive for a bit. Check out the bayou. Look at the Mississippi. Find the end of the Earth. He gives me the rundown on the status of installation of the advanced detector until we realize we lost the company of the Mississippi River when we missed our exit an hour back.

LLO is in the Livingston Parish, just a bit farther on past Baton Rouge. Architecturally, the two observatories, LHO in Washington and LLO in Louisiana, are about as similar as two buildings could possibly be. They are at first approach physically identical. Brian O'Reilly, the head of installation at LLO, tells me in an Irish brogue that they lock the opposite of the front double doors at LHO, and each time he goes to Hanford, he yanks the wrong one.

Culturally, there are palpable differences between the two

observatories. LLO has an undeniable southern lacquer. "This is Louisiana," they remind me. The reminder is proffered by an Argentinean, an Irishman, and an Australian. Typical of the academic side, the scientists are from anywhere. But the technicians, the control room operators, the support staff are, importantly, from Louisiana predominantly, and so the observatory is infused with a southern tinge.

The instruments themselves are fantastically complicated contraptions, too fantastically complicated to be identical, although Brian O'Reilly and Mike Landry as counterparts at the two observatories do try to minimize differences. There is constant communication across the continent, and the average quota of information they have to store, retrieve, assimilate, share seems inhuman. There's actuation systems and noise and seismic isolation and squeezed light and laser stabilization, mode cleaning, DC versus RF output, dark port versus light port, active vacuum systems, hydraulic systems, cooling systems, control systems. I wonder if there exists any one person in the collaboration who understands the whole thing. The individual previously deserving of that accolade is Stan Whitcomb. (Braginsky says of Stan, "Oh, a good fellow and an excellent experimentalist—a fantastically delicate person, gentle, wise, and very well educated and adept at public relations. A first-class experimentalist, no doubt.") Now Stan has other important work, like LIGO-India, which is exactly what you're thinking it is.

If you ask people to be brutally honest, to close their eyes and imagine the one person they would call now if the interferometer just would not work, I am told they'd say, "Rana." They would close their eyes and say "Rana," like an incantation. Don't be lulled by the lyrical Bengali name. Rana Adhikari grew up in Florida, his father a NASA engineer. He remembers playing in a Florida schoolyard in the sixth grade and turning skyward along with the other kids when the Space Shuttle Challenger seared the blue dome and rained back down as wreckage. His teacher cried as the kids, confused by the unidentifiable sparks, went back to their playground games.

I can't take my eyes off Rana. Jamie says, "It's a kind of celebrity." Part of Rana's charisma is related to the social power of indifference. He's not generally indifferent, though. He listens to other people sometimes with uninterest, true, and that could be interpreted, wrongly, as diffuse indifference. He might occasionally remark on someone's contribution to the conversation, usually with mild derision, his voice so smooth and calm you're expecting consensus, and then the meaning slowly resolves as mockery. When Rana infamously berates his charges, I bet his voice is exactly this honeyed and mellifluent, his criticism delivered without venom as though he regrets being the conduit for this unfortunate appraisal.

Part of Rana's social power lies in his seeming indifference specifically to external validation. He doesn't need for you to like him, self-esteem wise. I doubt this self-assuredness is humanly possible, but the impression, the illusion created is powerful. (Contrast: The first time I overheard Rana begin one of his stories with, "My friend Janna . . ." I burned with good sentiment.) In general, I imagine that when he gets older—he's currently in his thirties—he will seem sage, ancient, his humiliating assessments accepted as wisdoms to be received with gratitude.

Rana has a way of wooing the contraption, of negotiating the ifo back down. The interferometer talks to itself and gets into loops. And there are lots of channels for the machine to feed into itself. When I ask him about this reputation he agrees; he has a way with the instrument. He nods earnestly, not bragging, and explains he memorizes everything he can about each system so he can just think through issues and possibilities instead of trying to go back to a desk and a computer, paper and pens, to devote several hours to calculating. There just isn't the time for that sort of thing, and so he has to imagine it, imagine how it would all work, and then he can just sort of say yes or no, that will or won't work. He was worried he might not have that ability forever, but now that Advanced LIGO's installation is under way, he feels it coming back. Even the LIGO spokesperson, Gabriela González, says, "Yes, the machine just works much better when Rana is around."

I write him: "Rana, I want to go back to LLO in a couple weeks. Meet me?"

"Travel! I just got back from India and Australia. I will avoid planes forever . . . or at least until I forget about 16 hour flights. I'll check in with the Cajuns and get back to you."

It seems important to see the machine through Rana's eyes (big and dark and subcontinental—a powerful constant in a face he alters with ironic and brief mustaches, sometimes comically broad). When we make it to LLO, Rana asks about my last visit, "Did they tell you about the bass?" "No." "What the hell kind of tour did you get?"

The fresh LIGO arms encased in cement would sink into the swamp, so they had to dig underneath and sort of shore them up, leaving moats that run alongside the access road. The moats filled with water, again because of the swamp. Then something amazing happened. The moats became populated with bass. No one knows where they came from. I floated the theory that a storm swept fish up miles away, touching down with tornado-like delicacy, and then stumbled across the state to randomly throw them down on LIGO before either dissipating or gathering strength in the Gulf. I've adapted this theory from the film *Magnolia*. And I think I also read tales of cows deposited 3 miles south of their ranch on Tornado Alley, dead but otherwise unscathed. Rana thinks it's as good as the going favorite, which is that birds stick their feet in the mud and get fish eggs on them then they fly here and put their egg-laden feet in swampy trenches to spawn. I actually think this theory is better than mine on account of the fact that I haven't heard evidence of a rainfall of dead bass that missed the target moats.

Initially Rana didn't believe it about the bass. So one of the engineers caught one with his bare hands and brought the fish to Rana inside the lab, still alive, his finger in the gills, and Rana recoiled. "What the hell are you doing? Get that thing out of here." And the guy said, incredulous, "I'm going to put it back."

Rana tells me, "There are alligators. Did they tell you about the alligators? What the hell kind of tour did you get?"

After these stories I take note for the first time of a community corkboard in the hallway that features several pictures of men and women holding bass by the gills or posed on a slushy bank with an alligator partially submerged a few paces away.

Rana takes me to the roof of the building to get a better view. He says, "Do you see how all the trees in the one swatch out front are in regular formation? The Weyerhaeuser lumber company came and felled the trees and then planted these."

A lumber company knocking down a small wood in the front yard of the most seismically sensitive instrument ever built is hardly ideal. With the first-generation machines, after a year investigating, attaching seismometers along nearby Highway I-12, checking out industrial pipes in the vicinity, they couldn't identify the source of a particularly terrible noise. Rai drove to LLO at 6:00 AM, frustrated after a summer sunk in pursuit of the noise, a miserable realization washing over him as he watched trees fall on the land along the road to the site. Rai raced into the control room and sent the operator outside to watch. "Tell me when a tree falls," Rai instructed. "And boom, we saw it on the seismometers." During site selection, of course cognizant of the expansive Weyerhaeuser company, they somehow drastically underestimated the frequency of the logging. Rai offered to buy larger acreage from Weyerhaeuser to protect the experiment. The asking price too steep for consideration (possibly hundreds of millions?), they had to find a technological solution. The lab switched instead to an elaborate hydraulic system to isolate the mirrors. The active isolation system was already in the plans for the advanced machine but brought forward into the initial detectors to deal with the logging.

Generally LLO doesn't feel nearly as remote as Hanford. On the drive to the site, decrepit houses garnished with kids' toys splayed out front colonize a strip of quiet road, the colony embedded abruptly on the sparse side of a railroad crossing. The crossing gates are a kind of satire, a charade of a gated community. The decrepit houses give way to fully rotted houses. The splintered wood, once structural, now strains for the dirt, trying to reenter the life cycle, already accepting shrubs grown between shattered slats. Gaby González gives me a lift to the observatory in her convertible one morning and expresses some discomfort at the juxtaposition.

I say, "I heard the end stations got shot up," and Gaby says, encouraging me not to exaggerate, "Well, there may have been one bullet hole. And it was probably an accident."

I hadn't considered the possibility that it was intentional.

"Well, some people thought it was a warning shot or who knows what. But there was a meeting with the hunting association and that seems to have settled things. The hunters are aware there are scientists out here." She smiles at me reassuringly. I fake a smile in return. My fake smiles aren't worth the effort so I rarely don one, but I wanted to reassure her as she had me. I think we were roughly equally convinced by the other.

Back inside, there are more people than usual in the control room all in a crowd looking at the monitors. Today they opened the gate valve at the end station to the x-arm and they're shining the laser down the arm trying to hit the target 4 kilometers away. After several hours the target is hit, the monitor shows a pulsing blob of light in a very black frame. It's a big deal, but the atmosphere is quiet and low-key. Not a nail-biter. In a little under a year, they'll open the y-arm and try to hit the target with the laser and then both arms will be locked. That'll be the big event. That moment will be a proxy for flipping a switch, for turning on the machine. Champagne better flow. It's still months or even years from that moment until they're ready for a detection. The time will be spent trying to beat down noise, which is often unpredictable and resistant to control.

Inside the vast corner lab there's a fairly solid metal staircase to a short bridge so you can walk across one of the arms and then down a staircase to the big interior of the lab, interior to the vertices, a perimeter defined by the corner chambers and the origin of the two L-configured arms. A laboratory regular would skip the bridge, though. There are low metal surgical-style tables underneath the tubes to slither across. I want to ask how this convention developed but there isn't time before Brian O'Reilly is just a pair of kicking legs, shoe soles through paper covers, and then impish encouraging eyes under one of the tubes. There's no making it look cool, but I can see how after a while you just consider the options, trapped on the wrong side of two steel arms that break through the walls, and belly crawling starts to beat the staircase option after a few trips.

There are surprisingly few people in the cavernous interior to the vertices. Some are affixing cables, some are sitting underneath the tube near a gate valve and doing something, I don't know what, but I take note of their confidence. No one tells anyone else what to do. Everyone seems to understand the next step needed and seems expert. One person is in full bunny suit behind temporary clean-room drapes. He stands on top of a structure. I wonder, Is that my friend Aidan? He would be installing parts of the thermal compensation system, which adjusts for distortions of the mirror due to laser heating. But it's hard to make out individuals under a bunny suit and it's not like you can drop in and chat, so I fall onto my stomach and crawl to the civilization side of the arms.

Brian drives me to the end station and points past me out the side truck window to a small hunting hut, a modest wooden tree house on stilts, and a blue barrel to feed deer. Hunters hide out in the hut during the early hours of a new day and then shoot the deer.

"The bullet hole was not an accident," he says emphatically. "The FBI came and investigated, and that sort of put an end to the funny business," he concludes with a brisk nod. I believe him. No one smiles.

The observatory head in Louisiana, Joe Giaime, says, "To the Europeans we look ridiculous. Typically American. At one site a pickup truck drives into an arm and the other site gets shot up. All we need now is a hamburger incident." Still, LIGO is the only game in town—town being the Earth. (The European counterpart, Virgo, is not yet at advanced capability but aspires to join the advanced network soon.)

At the end station a cluster of cables hangs out of the side of the end chamber. Built-in cranes will lift the payload to populate the chamber, but the weight of the suspension system and mirrors is near the limit of the crane capacity, so some masses will have to be removed. Eventually people do have to go inside the chamber to finish assembly and affix connectors. The top of the chamber lifts off to allow entry, and there's very little room for a grown person to maneuver around the payload. It takes eight weeks to close a chamber after a cartridge touches down. When installation is complete and the humans and spiders have been removed, the end chamber will be pumped down to vacuum and the gate valve to the arm opened.

On our way back, Brian O'Reilly loops around to the y-arm and stops at the mid-station, where a construction crew rests on the access road, taking a break, masks hanging off necks and ears, legs stretched across the tar. They have been pulling insulation from the arm since early in the morning. There's also a small leak somewhere along the y-arm that has taken months to resolve, but they've honed in, and the insulation, infested with black widows and brown recluses, has to come off to finally get at it. Brian lets me step inside the cement encasement and see the continuation of the steel arm that we crawled under in the main building, but then he swiftly scoots me out due to the lack of a protective mask. The atmosphere of the tunnel is thick with mold. As we leave, one foot in the bright warmth of the Louisiana midday, the rest of me still in the thick dank dark of the tunnel, I can see down to a light at the far end, 2 kilometers away on either side. I imagine Rai Weiss walking the tunnels. He would have been the first person to walk along the arms inside the cement encasement in the years after its construction, his hand dragging along the steel spool, a flashlight catching confused vermin and snakes, an understanding forming quickly, an acceptance of the previously unconsidered forces of urine and chlorine on stainless steel tubes, cursing the 4 exceptional kilometers until he made it to the literal light at the end of the tunnel. I think, "I hope he wore a mask."

Construction of the two observatories began in the mid-1990s under the leadership of the second LIGO director, Barry Barish. "What am I going to do now?" the Caltech president asked the recently fired Robbie Vogt.

Vogt suggested, "They just canceled the Super Collider. Barry Barish is a particle physicist. He is damn good and he can run LIGO."

The Superconducting Super Collider aspired to be more than a multimillion-dollar hole in Waxahachie, Texas. The accelerator should have found the famed Higgs particle decades earlier if Congress had not terminated funding. Barry Barish was in charge of design of an experiment to place in the beam of the accelerator, but when the Super Collider was decommissioned in 1993, he barely had time to sink into disappointment. After the requisite due process on the part of the selection committees, the NSF, and the Caltech administration, Barry Barish had a month to respond to the offer to step in as director of LIGO. He made the decision overnight. He assures me that that is an exaggeration, but only a slight one. Barry had maintained an intellectual interest in LIGO since Kip first proposed an initiative in experimental gravity at Caltech in the late 1970s, and he let that idealism sway his decision. "All I had to assess was if I thought I could make a difference."

In 1994, Barish stepped in as director of a killed project, though not officially killed. A deathbed project. The money (not yet given final approval) was being taken off the table by the NSF due to a lack of confidence. Barish saw two immediate tasks. One was to build a team. The bigger task was to acquire the actual money, and by his assessment significantly more than had been requested. Not only was the NSF sitting on a project they were disposed to kill, but that project was going to require even more money. The lower budget was possibly a reflection of Robbie's skunkworks mode, while Barry intended to build a more robust management structure. Barish assessed the resources required as $300 million or more.

The budget created a tactical issue for Barish. He put the cost concern into conversations with the NSF immediately. The low fiscal estimate, if considered a problem of an earlier era, did not have to be the problem of the new era. If Barish had hesitated to broach the topic, there would be no LIGO.

In 1994, twenty-five years into the idea, Barish revived NSF confidence—"Kip wowed them and I brought in some reality"—with a revised budget that pushed up and over $300 million. (Kip counters the intended compliment. Much of the conversation was taken up with the prospect that the initial detectors would not detect gravitational waves, a prospect that Kip pressed as likely, and that an advanced generation of detectors would be necessary.) With funds not just promised but released, LIGO was able to escalate from a small group of innovative experimenters in comparatively modest campus laboratories to two enormous observatories maintained by scores of engineers and scientists. The facility, in 1991 a modest R&D control room housed in a trailer at Caltech and punctured by 40 meter tubes, had to upsurge a hundredfold, and twice: two warehouse-scale laboratories in Louisiana and Washington. Buildings had to be built, tunnels constructed, land procured. A volume of more than 18,000 cubic meters had to be pulled to a hard vacuum. Experts on precision measurements were drawn from neighboring fields; scientists and engineers were enlisted to design and build beam tubes and to oversee fabrication of lasers and mirrors. An expanding group labored in anticipation of the manufacture of real instruments on the real Earth with real detection capabilities. Kip said emphatically, "Barry Barish is the most skilled manager of large-scale projects we've ever had in the world." The view is widely held, possibly even held unanimously among those with an informed opinion.

Supplied with the information that Barish was born in Omaha, Nebraska, I interpret the thorough confidence, the belt buckle, the lanky stride, the rugged demeanor. He corrects my suppositions, curbing the cowboy imagery: "But I moved to California by the age of nine." Lean on pleasantries, he's still pleasant. There's an almost military candor to his parlance. He is neither loud nor soft-spoken. He earns admiration promptly, although as an adjunctive, not the objective. He's supremely good at making good decisions. Deft and effective, Barish forged not only the buildings and the instruments, but also a growing coalition of scientists. The LIGO scientific collaboration expanded to include theorists and complementary observers from around the world so that a community grew outside of the experimentalists, a global community invested in maximizing the astronomical payoff of the new observatories.

The problems of a large-scale gravitational-wave observatory were so new that they couldn't be solved in a conventional managerial way. The control system, as an example, had to be automated, a multidimensional interactive system of handles in a complex feedback loop that needed systematization in an analytic and reproducible way. There was a sense of awe at the Caltech 40 meter prototype for the guys who operated the controls seemingly by divination. An operation on a scale inflated by a factor of one hundred couldn't hinge on a clairvoyance acquired through the tweaking of the knobs. The control of operations had to be more robust, with convincing prospects for longevity. More than an interface issue, they needed deeper integration. Novel forefront designs had to fit and operate together. Barish acquired scientists from different fields, since no scientists were trained in gravitational waves. He hired control systems experts from the deactivated Superconducting Super Collider. Superposing them on the small LIGO contingent was problematic as resentments surfaced in those who no longer had their expert hands on the handles. (Jamie Rollins has since created Guardian, a sophisticated automation package to configure the control loops in the advanced instruments and keep the ifos locked in their most sensitive state.)

The industrial design was also pioneering. Although interferometers have a notable history in physics—most often cited is the Michelson-Morley interferometer of the 1800s that dispelled the fabled ether once (wrongly) believed to support light travel—there was no precedent for building a suspended mass ifo before Rai's first prototype in the Plywood Palace in the 1970s. There was certainly no precedent for extrapolation from the Caltech 40 meter prototype to a machine one hundred times bigger. Scientifically, the upscale had never been attempted.

They had a great project. They had money. They hired great people. Next they had to have the sites. "Once you get money from the NSF you want to show them you know how to spend it." He focused on breaking ground, building the civil side, the doors and buildings, the pipes and the vacuum system. The finer technology of the mirrors, lasers, and suspension systems could take longer.

The observatory in Louisiana occupies private land, which should have made the execution easier, circumventing the governmental red tape of the Hanford site. The plan was to build in Louisiana first. They broke ground by around 1996, but in Washington, not Louisiana.

In Louisiana there was some trouble with the Livingston Parish, which had a population at the time of around nine hundred. Louisiana is a right-to-work state, and they were picketed when they paved a 1.5 mile road to the observatory through public land. Other objections were more metaphysical. As LIGO held an open meeting for the Livingston citizens about the construction, across the street at a little school, fundamentalists met simultaneously and coincidentally to advocate for the teaching of creationism in the parish. A device to measure signals from a billion years ago seemed incompatible with their curricular ambitions. But there was some support for LIGO's efforts too. The first letter Barish received from a local resident came from a teacher in attendance at the meeting at the little school across the way. She implored him to bring the scientific campaign to Louisiana for the sake of her students. Barish, moved to consider the collateral impact of his work, the political and of all things the spiritual impact, understood that his sphere of influence had just expanded.

(Incidentally, nearly two decades later, there are still some suspicions about the lab. As they flew over the 4 kilometer L-shaped ifo on descent into Baton Rouge, a man apprised the occupant in the next seat on the plane, who happened to be a LIGO scientist, that the secret government facility they could see below them was designed for time travel. One of the arms brings you to the future, was the theory, the other sends you to the past.)

With some regret that the creationists might gain ground in the interim, and realistic about the greater difficulties involved moving policy versus earth, Barry quickly said, "We'll flip." Ground broke first at Hanford, which wasn't without its own issues. They had trouble mining water for construction, having to dig deeper than expected. The Department of Energy hesitated to approve the extended excavation. Barish suggests they didn't want them to find tritium or whatever else might be buried there on the site of the first production-scale nuclear reactors. "I'm not bad at twisting arms," he understates. Despite the varied and unique obstacles, the observatories were occupied and active by the turn of the twenty-first century. When the first LLO buildings acquired the aforementioned bullet holes (by some accounts there was more than one), the FBI effectively suggested they build a bunker around the experiment with high fences and other security measures. Instead, Barish went to lunch at the local hunting club. Soon the problems faded. But not before someone shot an alligator.

As for the global view, Barish may have sharpened the suggestion for a two-phase project: an initial detector, installed in the new facilities just after 2000, followed by construction of an advanced detector, installed as of the end of 2014, with early science runs expected in fall 2015. (The plan for initial detectors to be improved with more advanced detectors dates back to the original 1989 proposal.) The first stage would demonstrate the experimental potential, and there would be "possible" gravitational wave detections without breaking physical laws. (The first generation did demonstrate the technology but did not make a detection.) In the second stage, with the construction of Advanced LIGO, detections were "probable." (And here we are in anticipation.) Barry reflects, "As a scientist, you go into the unknown. As an experimentalist all you can do is achieve the experimental goal. Maybe Nature will be kind, maybe it won't . . . that's how you advance science."

Barish stepped down in 2005 to head the International Linear Collider, and Jay Marx was hired as the next director. Jay's job was to get the money for the advanced instrument currently under upgrade at LLO and LHO. Including initial LIGO, R&D, upgrades to the advanced machines, and operating budgets, the whole shebang integrates to about a billion dollars.

Jay Marx, in his current capacity as advisor, has lunch every week with David Reitze, the calm, easygoing, affable director since 2011. Reitze still enjoys the "fun stuff": the tinkering, the science, the experimentation. Barry Barish, Jay Marx, David Reitze—all of them credited as excellent directors, each in his era with different challenges to surmount—require less text than the previous years of risk and turmoil.

Today there is an international collaboration of more than a thousand scientists and engineers devoted to the endeavor in some capacity. There are similar but less powerful instruments around the world. A primarily Italian and French collaboration runs a smaller observatory, Virgo. There is a LIGO R&D facility in Germany (GEO). There are independent experiments in Japan (TAMA and more recently KAGRA). There is an initiative under way to build a third LIGO observatory in India, and that project has its own unusual challenges—clash of scientific cultures, geopolitical obstacles. This collection of acronyms, countries, and alliances should leave an impression of a massive international scientific collaboration, of big science. Of the planetwide network of detectors, LIGO is the most powerful.

Everyone is striving to get Advanced LIGO installed and calibrated and up and running. Nostalgia and sentiment and the base-ten number system have highlighted a deadline: the hundredth anniversary of Einstein's paper on gravitational waves.

As Rai puts it, "We have to keep working if we're going to make a detection by 2016, which I think is absolutely essential, because I want it to be. I want the centennial. That's my mantra, that we should detect by the centennial of Einstein's paper.

"That would be a nice closure to this whole damn business."

15

Little Cave on Figueroa

In Highland Park, a Los Angeles neighborhood not far from Caltech, some of the research scientists go drink on Tuesdays, as the text that invited me put it. There's a complete ban on pretension, especially within certain ranks in the physics subculture. There's no dressing up facts or glamorizing. Use words as sparingly as possible to cut as close to the blunt data as possible. "We go drink on Tuesdays. You can come if you want," the message read.

Little Cave on Figueroa has a late happy hour and free tacos and a space that is marginally "outside," despite the brick walls and roof overhead, where people drink and smoke. Yes, smoke. The smoking is surprising because who does that anymore? Nobody smokes. Europeans do I guess, so part of the rampant smoking quantifies the European demographic, although the Americans partake too, but it seems to me to be with less vigor. Less conviction.

Little Cave is darker than standard bar etiquette requires and the music is cool and the bartenders look punk in a 1980s sort of way, which would be terrible if they had lived through the '80s, but instead they manage to land the ironic target. So I like Little Cave on Figueroa despite the fifty-minute walk to the train, despite computer equipment that is never light enough, despite the injurious weight of physics papers, overnight bag, gloves, and sweaters in preparation for the odious descent in ambient temperature through LA nights.

Clustered around a table designed for two are many bar stools occupied by a collection of scientists with accented English. We're transients. There are always some discussions about trajectories—"Were you at MIT when Rana was there?" Someone has always just arrived for a two- or three-year research post, for graduate school, for the very rare and coveted faculty position. Someone is always leaving for a two- or three-year research post, for graduate school, for the very rare and coveted faculty position. Generally, faculty are not invited to drink on Tuesdays.

I'm infiltrating the experimentalists' ranks. I have questions. Genuine questions that are not taxing and are not intended as tests of anyone's competence. They are the experts on the instrument. I'm the outsider. So I'm glad when the initial curiosity over my attendance on the best night of the week, Taco Tuesdays, subsides—Jamie says, in an undertone, "You're a scientific dignitary," I hope without sarcasm—and the drinks flow and inappropriate stories are told and I become one of the guys.

The postdocs—shorthand for postdoctoral research scientists—are particularly transitory. Their apartments are multimedia installations, explorations of the notion of transitory. I've been in some of their places, and there are internationally traveled childhood relics, nondescript couches found near college campuses more than a decade earlier. Brutally worn out furniture that doesn't quite fit in the space expresses a general lack of interest in the actual architecture. Built-in fireplaces are covered by bikes. In sum total the apartments say, "I don't plan on being here forever," even when the resident unexpectedly stays in that noncommittal arrangement for years. Two or four or five, traveling from that base, never quite settling in, never quite committing.

The connections between the scientists will last for decades, for lifetimes. They will meet here at Little Cave, or in Louisiana at the LIGO facility, or in Italy at the European Virgo, or in Nice for the next collaboration meeting. In Japan and in India we will see one another, and the continuity of our lives floats in that stratum, in the layer above the ground, in our heads, in our collaborations.

Scientists are like those levers or knobs or those boulders helpfully screwed into a climbing wall. Like the wall is some cemented material made by mixing knowledge, which is a purely human construct, with reality, which we can only access through the filter of our minds. There's an important pursuit of objectivity in science and nature and mathematics, but still the only way up the wall is through the individual people, and they come in specifics—the French guy, the German guy, the American girl. So the climb is personal, a truly human endeavor, and the real expedition pixelates into individuals, not Platonic forms. In the end it's personal, as much as we want to believe it's objective.

The membership at Taco Tuesdays is selected on the basis of friendship as well as scientific interest, and the warmth among the people who meet here is real and unprofessional, as is the conversation. There's also a certain level of exhaustion.

I first came to drink with the postdocs in February 2013, and LIGO was maybe somewhere in the middle of its upgrade to advanced capability. It will effectively be a new instrument known as aLIGO, a shorthand for Advanced LIGO. I say something to the gathering about 2015, a projection for the first direct detection, and there are chuckles—it wouldn't be accurate to say bitter chuckles—and lots of hanging, shaking heads. No way. No way. Not 2015. "Well, maybe," someone shouts. Maybe for a month or two for tests but no. No detections. Some very pessimistically suggest 2018. Tonight, late March 2015, they may have reason to be more optimistic.

These are people born in the 1970s or 1980s. Maybe one or two in the 1990s. None of them knew Joe Weber. But it's like they're building the same ship. Or wait, mix the metaphors, it's like they're going for the same treasure. Reading from the same map, picking up where the steam-punk technology failed—an insane, doomed, impossible bar detector designed by the old mad guy, crude laboratory-scale slabs of metal that inspired and encouraged his anguished claims of discovery.

The collective exhaustion is further degraded by nagging uncertainty, each person distracted by silent concerns over the different domains of responsibility (the suspension system, the optical fibers, the transition to DC output), the group distracted by audible concerns over the shared project (the urgency of installation, the sensitivity goals, community opinion). But they are also kept afloat by nervous excitement. The machines are nearly operational, and every day they sweat to haul the mission tangibly closer to actualization.

We've all developed stylized ways to describe the impending discovery. We all have our own idiosyncrasies in presentation, in the choices we make to avoid technical jargon and skirt concepts that require too much heavy lifting. I've listened to countless variants. I listen to several abbreviated versions tonight. The men and women around this table are part of the team that has won the hard-earned right to be part of that discovery. The search is not merely a search for black holes. The quest is not an inventory errand, an indexing of known objects.

We'll be listening to the direct messages of a fundamental force, brought directly to us by the fundamental force carriers. We'll be listening directly to the messengers of a fundamental law of nature. I'm running out of new amalgams of "direct," "message," and "fundamental," but you get the meme, as does everyone at the table tonight, regardless of which combinatorics they strike in their version of the yarn.

Only messages sent by the gravest, most cataclysmic concentrations of gravitational effort will reach these machines. That puts bare black holes, the big bang, and exploding stars in a privileged position. So while our ambitions are broad—a direct commune with fundamental law—we can also indulge in our admiration for the individual creations that inhabit this terrain.

When black holes collide the space around them rings until a perfect, spinning, bigger black hole remains and the space goes quiet. All compact binary mergers whir up in pitch and decibels rising to a characteristic chirp. The details of the orbits modulate the sound so we can reconstruct the path of the mallets ringing the drum.

When neutron stars collide they very probably make a black hole, although chunks of neutron star crust can slough off in the process, lightening the load enough that possibly another neutron star remnant settles after the commotion. The neutron stars are essentially undetectable by telescopes until the merger. But on impact (loosely defined) the magnetized, superconducting, condensed nuclear globes shatter to release a blast of gamma rays (higher-energy light than X-rays). A category of known, observed, and studied gamma-ray bursts (shorthand: GRBs) have been attributed to neutron star collisions.

Satellites have seen bursts. Pictures have been taken, though not resolved. The satellites cannot focus to a fine-grained image of the explosion, which lasts for only a fraction of a second. But they can track the emission of energy, see the burst spike then dissipate and sometimes document a paler afterglow. The collaboration between the gravitational observatories and the satellites that collect light significantly advances the scientific prospects. LIGO can record the final minutes of the inspiral and trigger the participating satellites to redirect their sights and search for an imminent burst. (Also, the reverse works too, since LIGO retains the data to be analyzed after the fact.) This developing area of research goes by the name "multimessenger astronomy," since the data comes both through light waves and through gravitational waves.

The supernova event that formed the compact remnants is another potential candidate. Every few centuries, a star explodes sufficiently near to us that we can see the thing with our simple human eyes, no intervening telescopic lens required. Typically a star blows every few hundred years in our own Milky Way, but with much less power in gravitational waves than black hole collisions. Even Advanced LIGO will have trouble hearing supernovae outside of our galaxy if current theories are right.

An exploding supernova sounds distinct but depends on the details of the detonation. It moans like a whale or cracks like a whip. The sound is a direct reflection of the accelerations of mass during the calamity. All supernovae are characterized as bursts, and there is a subgroup within LIGO devoted solely to the detection and analysis of bursts, both foreseen and unanticipated. Although there are some proponents gambling on supernovae as contenders for a first detection, many more suspect they will be too quiet to pick up with any regularity.

An isolated spinning neutron star is another archetypical

source of gravitational waves. If the surface of the star is perfectly round there are no waves in the curves of spacetime. But any mountain on that surface puddles the shape of space with each rotation, like a paddle swirled around in a lopsided loop. The sound from the spinning slightly mountainous neutron star is a pure unmodulated tone. It doesn't get louder or change in pitch. The spinning imperfect neutron star generates a continuous monotone.

The big bang was likely a cacophonous, chaotic mess. The gravitational noise of the universe's creation should average to a featureless white noise, a hissy static—by now, nearly 14 billion years later, a very quiet hiss. From our current understanding of the evolution of the universe immediately after the big bang, that noise is stretched to near silence with an inflation of spacetime within the first trillionth of a trillionth of a trillionth of a second. But, yeah, the big bang made a bang. There's no expectation that LIGO would hear the earliest moments since the gravitational waves are too quiet by now. While they're below the LIGO range, ifos in space could directly detect the remnant sounds of the big bang in several decades if those missions succeed.

Other possibly stochastic sounds could come from uncorrelated compact objects in different galaxies just howling incoherently at our detectors. An overlap of compact binaries could create a stochastic background, but that might not be a terrible problem until interferometers make it to space.

When I first heard Kip lecture about the prospects for a new window on the universe, I was hoping for the unforeseen, the unanticipated. Are there astrophysical phenomena we have not even imagined yet? Can we hear dark matter? Dark energy? Dark dimensions?

At the end of a long day the conversational interludes favor the specifics of the experimental challenges. Noise is a big topic. There are people in the collaboration who work just on noise. Noise is a factor in absolutely any scientific experiment and is in no way special to this experiment, the terminology not intended to play on the idea of gravitational waves as sound. Noise just means errors in any detector's ability to pick up the mark. In gravitational-wave experiments, noise can do both jobs. It can mean the errors you have to tolerate, your lack of precision. It can also in this particular industry reference sound. Listen to a conversation at Little Cave on Figueroa and pick up the significant background sounds mixed in with a friend's words. The signal I want is my friend's voice, but she's drowned out by the music. There are elaborate algorithms that people write to remove predictable noise, like the music in our bar analogy, but it's hard to exclude all of the other voices, to hear the one you want to resolve and to subtract the others. Quite possibly all the gravitational waves detected by this machine will be quieter than the background noise. The data analysts have to pick out particular sounds that are not even loud enough to peak above the din, the sound buried in the noise.

If the gravitational wave recordings can be linked to a bright luminous source in the sky, the corroboration of evidence will be far more compelling than a recording alone. Skeptics like Ostriker might require a multimessenger portrait before conceding a wager.

The younger generations of scientists rotate in and out, not egoless, but as though themselves components in an expansive machine. Installation of the advanced mirrors, lasers, and seismic isolation systems is complete at both observatories. The next phase is commissioning, which amounts to integration of the installed subsystems to create a functioning whole. The ifo at LLO has been locked, and just the other day the ifo at LHO was locked too. The recently installed advanced mechanism was brought into operating mode by small groups of women and men who fluxed over the weeks, each experimentalist working with whoever else was there on rotation, so that no one person gets the praise. I check the logs at night sometimes and was kicked into insomnia by a log posted at 4:23 AM: "We had another lock of about 40 seconds, during which all the signals seemed more stable than last night." There are setbacks: "Today was a particularly bad day for the whole locking business." Then a couple of nights later, at 5:24 AM, "We have been locked for more than 1 hour with CARM controlled by digital REFL91 (at 0 pm offset) and DARM controlled by ASAIR 45Q. The power buildup is 1100 times the single-arm buildup, giving an interferometer recycling gain of 33 W/W. . . . This means the interferometer visibility is about 94%." I consult a glossary of acronyms to better understand, but the pleasure is immediate. The ifo is locked and the sensitivity, though not yet good enough, is good.

The next morning the log is dotted with congratulations and words of thanks to the team of scientists who now drive the machine through to completion. Rai posts, "The first noise spectrum! Nice going." Initial LIGO required nearly four years of tuning from the moment of first lock before design sensitivity was achieved. Advanced LIGO seems to be moving much more quickly. The projection I hear tonight is that the next six months will be burnt improving the machine sensitivity until its response is delicate enough to hear faint echoes. Then the first science runs will begin in September 2015, during which the machine will need to maintain lock for several weeks. They aim to measure the difference between 4 kilometers and 4 kilometers plus or minus one ten-thousandth of the nucleus of an atom. The current director, David Reitze, posts to the log, "Fantastic! Noise hunting licenses for everyone!"

Many of the scientists at Little Cave on Figueroa do stints in the control room and the Laser and Vacuum Equipment Area. They'll write the control system codes, test mirror coatings, solder electrical pieces. They come to Caltech to visit or to live, if temporarily. Our conversations around the table are a mashup. We are ordinary and mundane one sentence and entirely abstract and ungrounded the next. We mix topics, tease one another, banter, flirt, and inject technical jargon.

The last glass is drained. We break out into the midnight color of a stark street. Raised arms are waved in vague directions. Pairs branch out, peel away along the sidewalk, and cross streets. We head back to ramshackle homes and student-quality sheets, to shared beds and friends' couches. Arguments pause to be resumed tomorrow. The noise of the bar perseveres but vaguely like the tone of a tuning fork. The tinnitus, mercifully temporary, is audible only when we strain for silence, only when we surrender at staggered hours to the near quiet of our private, woozy thoughts.

16

The Race Is On

Robbie Vogt summarizes, "They will definitely see gravitational waves, no doubt about it, but it will not be my discovery. I will read about it in the newspaper.

"I have no regrets. None. . . . The wounds are healed. This is ancient history for me now. I have a new career. I work in national security. . . . I'm not an employee of anybody. I'm a free agent. I pick the work I do. But I'm not an employee of the government. When I give a debriefing in Washington to the executive branch, I can say things that admirals and generals cannot say. . . . I'm paying back because I have the freedom to speak my mind . . . and it's a damn exciting career for an eighty-five-year-old man.

"Caltech was my country. It was something I could relate to. It was my family and my country. It is sad that I'm no longer part of Caltech now. . . . I've had many setbacks but every setback, somehow life compensated me again for it in some way. At the moment I got fired or if I had to step down or . . . it was very hurtful. It was devastating . . . but there were always people at the right moment who gave me an assist. Every time there was a change in my life, there was always a human being involved to help me. I was lucky.

"I happen to be a person who believes in nuclear disarmament and I work in nuclear weapons, among other things. I make it possible if this country and the world ever wants to disarm. . . . We were against proliferation . . . I never advocate zero . . . I always advocate a few. Because it will never be zero. It cannot be zero because people will not trust each other . . . but if you reduce it to a few dozen, people can no longer destroy the Earth with them. Right now, with four thousand you can make the planet uninhabitable, and I am afraid there are crazy people in the world who could start a nuclear war. But if they don't have four thousand weapons, if they only have twenty-four, then we will destroy a city, that's not the end. Not good, but it's not the end. With four thousand it's the end of life on Earth. And it will take generations of miserable deaths and I want to prevent that. I am most effective as an insider. Everyone knows I'm against nuclear weapons. I fight for the things I believe in."

Two months prior, Vogt canceled our meeting on short notice, deferring to health issues. I catch a rumor that he was in Afghanistan and his convoy was ambushed. He was the target of a retaliation for his weapons work. He was injured in the attack and has required repeated and unsuccessful surgeries— something near his spine, a fragment?

"I owe this country. This country has been kind to me. Much kinder than the country I was born in."

We stand outside the LIGO building and say a very protracted good-bye. We stand and shift feet. People fall out of the old wooden doors and look twice. LIGO scientists wave but don't say a word, glancing at Vogt. He wants to talk. He wants to talk about the work he does and why, about his fears for this country. He doesn't need my affirmation or my agreement. I don't offer it. I don't express my opinions on disarmament or Afghanistan, and they're not relevant here. I just listen. There wasn't a cursory word spoken. I am almost winded at the end of our many hours in unremitting conversation. I observe that I have no judgment against this controversial man. I don't express any political attitudes of my own, even if his are not mine. (This is unusual for me maybe.) And then I wonder, uselessly, if this meeting has prompted an NSA reading of my e-mail.

Vogt stayed in the LIGO collaboration in another capacity for a couple of years before he quit, although he may have felt he had been fired. Barish did not refuse the resignation. Barish says, "He just couldn't be in a non-leader role. I don't know that I could either." Rai hopes to reconcile. He and the three subsequent directors strategize to bring Vogt to one of the sites, to convey appreciation for the role he played, to edit that complex chapter.

Over dinner near LHO, the two of us in a booth big enough for six, Rai recounts the last time he saw Ron Drever. When Barish stepped in as director, he dissolved any lingering prohibitions against Ron and encouraged him to join the wider LIGO effort, hoping to dispel any enmity. Drever joined the LIGO Scientific Collaboration, attended the meetings, and continued to think of ways he might contribute from his own laboratory. He was often quiet, observing, but with no animosity, as though attending a friendly outing.

At a LIGO Collaboration meeting in Pasadena in spring 2008, Rai noted Ron's absence. He was concerned to learn that no one had seen him for a while. An uneasy Rai went to Drever's Caltech apartment. He opened the door to an unkempt mess of books and rumpled clothes. In the shambles, they found a little cubbyhole where they could sit together, Ron in an easy chair,

Rai in a hard chair. Somehow these details matter. As always they chatted about LIGO. Rai reported on the poor health of a colleague from Scotland. Then an hour into the conversation, asking again after his colleague from Scotland, Ron received the bad news as though for the first time, became worried afresh. Rai worried too.

Ron, confused and forgetful, refused to see a doctor despite Rai's urging. He complained to Rai that doctors are expensive. Rai stewed. "The guy is completely alone in this country. He never got married. He doesn't have any friends. He's in this disheveled apartment. And he doesn't come to work anymore."

Goldreich tells me the last time he saw Ron Drever, "I had to put him on a plane in the end, send him home to his brother. He has dementia now." He mutters this last bit, regretting telling me or regretting the fact of it. "I bought his ticket. I got on the plane with him. Took him as far as JFK. Got him on his next plane to his brother. It's sad."

When asked in 1997 his thoughts on LIGO—the full-scale instruments had not yet been built—Ron said it could go either way. He imagined that people will either view LIGO as a huge success or a total waste of money.

During the Drever affair, Kip maintained a friendly if strained connection with him. His respect for Ron's technical ability never faded. (Rai relays a representative incident. Kip was neck-deep in a tedious, multipage calculation when Ron presented a diagrammatic solution. Ron could not have performed the formal mathematical calculation but still he could somehow see the solution in pictures, impressing Kip indelibly.) He attended Kip's group meetings until his health wavered and his disorientation became noticeable. Despite the confrontations over the years, Kip feels on a good personal basis with all of them. He isn't on campus that often, nor is Robbie Vogt, but they bump into each other occasionally and will stand in the warm walkways and chat.

Kip always expected that LIGO would be a success. He reflects on the decades of technological challenges and the unpredicted political and psychological hurdles they overcame along the way and he's "amazed." He always suspected success, but in the early days, he hadn't fully anticipated arduousness. He's proud clearly of the experimentalists, beaming with admiration for the congregation that has fulfilled, at least technologically if not scientifically, the vision he conveyed with some prescience more than four decades ago. Kip, with his students, invested years of effort quantifying sources of noise, and he even contributed to the nuts and bolts of the buildup with an analysis of scattered light in the beam tubes, which helped shape the specs of the instrument. But his talents and desires were always more toward the theoretical, and sometimes the speculative. Kip's most valuable contribution to LIGO, by his own assessment, was to formulate—"in consultation with many colleagues and students"—the vision for the scientific potential of the experiment. He was relieved when the team was finally sufficiently robust in areas where Kip had expertise so that he could return to pure theoretical predictions of the sounds of the sources. As his last major contribution to LIGO, Kip wrote the scientific case for the advanced generation of detectors. "I'm happy to be watching from the sidelines now," checking the latest sensitivity curves only every few months. He acknowledges his good fortune in his new career in the film industry writing treatments for blockbusters (*Interstellar*), producing movies, attending premiers with his friend Hawking.

Ron Drever is alive but extremely unwell. His brother writes to me, "My mind is filled with so much about Ronald, who remains just remarkable. I visited him yesterday in his care home, where he has been for some two years; the caring is totally excellent. I cannot be sure that he really knew me, but probably." Joe Weber is gone. Robbie Vogt has never been to the LIGO sites, although every director has invited him. Rai is under the hood and Kip checks in regularly. Of the original German group, Billings is 101 at the time of this writing. Braginsky combats faltering health, striving to remain active until Advanced LIGO detects its first gravitational waves. His group continues to play significant roles in the technology development. Stan Whitcomb is securing a LIGO observatory in India. Jim Hough from Glasgow supplies essential components to the advanced detectors. He tells me, "We're all just trying to stay alive."

Rai says, "It's going to be hard. I wish I didn't feel this way." In July he'll troubleshoot problems with the mirrors at LLO that currently hang up the progression toward lower noise. In August he'll head to LHO to measure some nonlinearities in the digital-to-analog converters that drive the test-mass controllers. "I feel honor bound, duty bound. This system has to make a detection of some sort or we've led this country down a garden path." Although he's pushing for results by the centenary of Einstein's first paper on gravitational waves, published in 1916, if they miss that deadline, he'll accept instead the centenary of Einstein's 1918 paper on gravitational waves. "I don't like it. But okay." The original Einstein paper has errors anyway, he reminds me.

The first science runs had better make a detection, any detection, a first recording of the sounds from space. "Goddamn. It has to work. But that isn't really what I'm after, I hate to tell you that. If we don't detect strong-field gravity, then the thing is a failure. . . . We've got to detect black holes. That would be a satisfaction beyond belief. That would be a great thing, why all of this was worth it."

Rai spent the past six months thinking over prospects for instruments beyond LIGO, brainstorming with young MIT scientists—Lisa Barsotti, Matthew Evans, Nergis Mavalvala—who also have their sights on the far future. Quantum experiments are under development. There are discussions of 40 kilometer machines. There is a mission proposed to launch ifos in space. There's vigor in his description of his latest ideas. He stresses that the collaboration must continue to move forward. They need to design next-generation instruments now, not after detection. That will be too late. The field will stall. He conjectures about a far future of high fidelity, his one ambition fulfilled—not the hissy crackle of this generation's anticipated detections, but instead an unbelievable sound coming from the speakers of an incomparable recording device. He says, "It won't be in my lifetime, but that's not important."

Somewhere in the universe two black holes collide, an event as powerful as any since the origin of the universe, outputting more than a trillion times the power of a billion Suns. That profusion of energy emanates from the coalescing holes in a purely gravitational form, as waves in the shape of spacetime, as gravitational waves.

The first gravitational-wave train humans aspire to record is at this moment in a race against the completion of the Advanced LIGO machines. Initiated by a collision of black holes or neutron stars or exploding stars, maybe more than a billion years ago, the waves in the shape of space have been on their way here ever since.

A vestige of the noise of the crash has been on the way to us since early multicelled organisms fossilized in supercontinents on a still dynamic Earth. When the sound moved through our Local Supercluster of galaxies, dinosaurs roamed the planet. As it passed the nearby Andromeda galaxy, the Ice Age began. As it entered the halo of our Milky Way, we were painting caves. As the wave approached a nearby star cluster, we were in the final furlong, the rapid years of industrialization. The steam engine was invented and Albert Einstein theorized on the existence of gravitational waves. When I started to write this book, the sound reached Alpha Centauri.

In the final minuscule fraction of that billion-year journey, a team of hundreds of scientists will have built an observatory to record the first notes from space. As the sound moves through the interstellar space outside the solar system, the detectors will be operational.

As the wave nears the orbit of Neptune, we have only a few more hours. Past the Sun, we have eight more minutes. Someone will be on duty in the control room, awash in fluorescent lights, listening to the detector through conventional speaker systems or headphones for fun, because she can. And maybe beneath the noise of the computers, the fans, the clack of computer keys, the noise of the machine itself, after the passage of eight unexceptional minutes spent fidgeting with the control system, she might barely hear something that sounds different. A sophisticated computer algorithm will parse the data stream in real time and send a notification to the data analysts— preferably in the middle of the night, triggering a fumble for glasses or a stumble out of bed for dramatic effect—and someone will be the first to look over the specs of the trigger and think calmly, "This might be It."

As much as this book is a chronicle of gravitational waves—a sonic record of the history of the universe, a soundtrack to match the silent movie—it is a tribute to a quixotic, epic, harrowing experimental endeavor, a tribute to a fool's ambition.

Epilogue

On Monday, September 14, 2015, the sites weren't quite ready. The first advanced science run, O1, was postponed by a week from the scheduled start time of 8:00 AM that morning. In a science run, the ifos would be locked without intentional intervention, the machines left undisturbed to collect data. Instead they extended ER8, engineering run 8, to test systems, implement last-minute tweaks. The priority during the engineering run was to improve stability, keep the machines locked, and get the alert triggers ready. Algorithmic pipelines automate a level of data analysis, searching for worthy signals in the data stream. The procedures weren't yet in place for the pipelines to alert observing partners, teams that operate telescopes and satellites that might respond to the alarm and search for a counterpart in light. The machines were ready, but the pipelines were only partially operational, so they allotted themselves an extra week to collect data, but not preciously, permitting plenty of disruptions and interruptions.

The season is windy and microseism is high due to storms coming from the direction of the Aleutian Islands, or the Gulf, or Labrador off the coast of Canada. The weather beats against the continental shelf and the seismic activity can kick the machines out of lock. Both sites were having trouble. LHO achieved lock early in the evening on Sunday, September 13. A graduate student performed some tests late Sunday night, finishing up Monday morning at 1:00 AM. Rai was at LLO over the weekend, fretting over a source of radio noise but reports, "Luckily, my wife said I had to get home." People continued to run tests at LLO but quit in frustration in the darkest hours of Monday morning, the ifo finally locked.

There's a window of under an hour when the machines were left in observing mode undisturbed. At 2:50 AM at LHO and 4:50 AM at LLO, both detectors recorded a burst. There was one operator on duty alone in the control room of each facility, but neither could have heard anything. The signal would have been too brief to parse by ear.

An automated pipeline found the event within 300 seconds of recording the data and silently documented its noteworthiness. Potential candidates are often reported, so there was no drama when people in Europe woke up and checked the logs, as they would habitually, and saw the notice of a candidate event. Calls were calmly placed to the sites to check the status of the ifos. The two operators confirmed: all was well.

They froze the instrument and collected background noise until lock was lost and the ifos dropped offline at both sites within a few hours. By the time Mike Landry began his day and checked the logs, they were full of traffic and conjectures about the candidate. Immediately Mike thought, "It's a blind injection," a false signal intentionally but secretly injected into the data stream to test the collaborations preparedness and ability to cope with a bona fide signal. Somewhat frustrated with the blind injection team, composed of three scientists selected to perform the tests, Mike thought, "What are they doing? We're not ready yet." He drove to LHO to attend the usual 8:30 AM Monday-morning meeting. One of the blind injectors happened to be on site. Mike took the opportunity to inquire patiently, "Are we in a blind injection phase?" Respecting the rules that ensure blindness, he cannot ask explicitly if there was an injection. The team would refuse to confirm or deny. But he can ask if they're in a phase of performing blind injections. His colleague responded, a little flustered, "No."

"Did you test blind injections?"

"No."

"Did you test regular injections?"

"No."

Maybe he just wasn't phrasing the question quite right. After several other permutations, he asked, "Did you test any injections of any type?"

"No."

Mike thought, "This is not a drill." He tells me, "After I realized there was no injection, I became very cold."

By 9:00 AM Mike joined the main weekly telecon with the international collaboration. There were lots of voices on the line, probably most conjecturing, as he had, that this was a drill. Jamie Rollins says, "I was utterly incredulous." Mike took the duration of the call standing up, simultaneously trying to reach Gaby González at LLO. He finally said to those on the telecon, "This was not a blind injection." The voice of Alan Weinstein from Caltech came on the line: "Mike . . . can you say that again?"

. . .

In mid-December 2015, I receive an e-mail from David Reitze, the director of LIGO. The subject line: "CONFIDENTIAL communication about LIGO." The message reads, "On September 14, the two LIGO interferometers recorded a signal consistent with the inspiral and merger of two ~ 30 solar mass black holes." Dave continues, "Over the past three months, the LSC and Virgo have carefully vetted the signal and have concluded definitively that we have made the first direct measurement of a gravitational wave and observed the first binary black hole system." The letter is signed Dave, Rai, and Kip. "We all stress that no information about the detection can be made public until the paper has been published, likely sometime in February." I don't want to tell anyone. I'm rattled. I spend the next few hours silently, trying to imagine the event, trying to visualize the bare black holes colliding, shaking spacetime, sending the noise our way—trying to believe viscerally.

The collision sent us the most powerful single event we have ever detected since the big bang, the power in gravitational waves a hundred billion trillion times the luminosity of the sun. The detectors caught the final four orbits of a black hole 29 times the mass of the Sun in a pair with a black hole 36 times the mass of the Sun. Only a few hundred kilometers apart, the holes circled very nearly at the speed of light. As they fell together, the event horizons, distorted by their proximity, collided and merged, shedding imperfections to ring down into a quiet black hole more than 60 times the mass of the sun. The recorded signal of those final few orbits, the collision, and the ring down lasts 200 milliseconds. The ifos detected changes across their 4 kilometers about a ten-thousandth of the width of a proton, displacements right in the range Kip and others had theorized decades before. The event is still considered loud, spiking above the background noise in places. The signal can be sonified, but the recording has to be slowed down for a person to discern the structure, the mounting pitch of the chirp as the black holes draw together, the descent of the formation of the final coalesced black hole. There are other bumps in the data, but none so glaring and clear and distinct. The rarity of such a loud event is difficult to ascertain.

With this one detection, LIGO hit the centenary of the general theory. Einstein first presented the geometric description of gravitation on November 25, 1915. Strictly speaking, the collaboration beat Rai's target, which was the publication of Einstein's subsequent paper on gravitational waves.

Rai says, "The main thing is that I have this monkey off my back."

He teases, "My wife is suddenly very interested in this field." By coincidence, Rai's longtime friend from the NSF, Rich Isaacson, came to visit him in Maine that week in mid-September. Rich's first reaction to the news was endemic doubt. "Do you believe this?" he asked. There were even concerns Rai shared that a signal could be injected maliciously, by hackers, but the level of detailed knowledge such a hack would require likely exceeds any one person's, except the most able and involved scientists within the collaboration. And those few were interrogated as a matter of precaution. Disbelief reluctantly gave way to a hesitant excitement.

When I catch up with Kip, he says, "I had a moment of profound satisfaction." He always expected black holes would be the first sources detected, regardless of fluxing attitudes that dominated community opinion over the years. The more massive the black holes, the louder the collision. They can be heard from farther away, which makes more of them available to the detectors, despite their intrinsically smaller population. To Kip, the question was: When?

I relay to Kip that the experimentalists advised me to be patient, believing the first detection would not come for years. "Well, not Rana," he corrects. "Rana kept telling me there'd be a detection this time."

I ask Rana about his prescience. He replies, "I always say that."

Still, when he heard talk in the halls about the candidate, Rana was unmoved. "We just turned the thing on." Unflustered, he took a day before finding the chance to look at the data and thought the perfect simplicity of the signal almost absurd. There are no prominent deviations from the wave predicted theoretically. (In a few minutes, using my own black hole codes, I reproduced a template that looks very much like the waveform from the final orbits and chirp of the cleaned-up data.) He hoped for a challenge to general relativity. He hoped he had helped build a machine to test quantum gravity. "We're just going to have to work harder," he says.

Rai gives himself exactly one minute's reverie, "Black holes. That's what all the old-timers were after. Pure geometry. Pure spacetime coalescence." But then he worries about the future. The detectors are more sensitive already than the initial generation. The comparison can be put in perspective this way, quoting from a draft of the publication: "The 384-hour period reported here surpasses all previous observations of black hole binaries combined." But the ifos' advanced sensitivity should be even better. Rai is back at work, helping to drive the noise in the detector down further, to lean on the instrumentation harder. The team has to labor still in order to operate a fully productive observatory as promised.

I say, "Congratulations, Rai, I cannot express my excitement. I cannot imagine yours."

"Well, the monkey is off my back, but now the monkey is walking alongside me. Ask the monkey."

Two very big stars lived in orbit around each other several billion years ago. Maybe there were planets around them, although the two-star system might have been too unstable or too simple in composition to accommodate planets. Eventually one star died, and then the other, and two black holes formed. They orbited in darkness, probably for billions of years before that final 200 milliseconds when the black holes collided and merged, launching their loudest gravitational wave train into the universe.

The sound traveled to us from 1.4 billion light-years away. One billion four hundred million light-years. A few hours before the wave hits the Earth, LHO is locked. An hour before the wave hits, LLO is locked. In the middle of the night, the scientists in Washington abandon the long day's efforts and head home. Scientists in Louisiana put down their tools in frustration to leave the instrument undisturbed in observing mode. Within the hour, a signal sloshes over the Earth. The gravitational wave came from the southern sky, skimmed past Louisiana, ringing LLO first before cruising at the speed of light nearly in the plane of the continent to hit LHO 10 milliseconds later.

By 8:00 A M, the wave is already past the Earth by two billion kilometers. Rai, on vacation in Maine, checks the logs as he always does, to see if there's something needed, some way he could help. There he sees entries in red about the freeze placed on all activity at the sites. Rai begins to wonder, along with the rest of the team, "What the hell is going on?"

There will be press releases. Scientific papers will be published. News articles will abound. There will be documents, reports to the NSF, and careful records kept. We'll continue to send a portrait of our best selves into space, a declaration that we were here, that we strove for understanding, that we often

failed and occasionally succeeded. We heard black holes collide. We'll point to where the sound might have come from,
to the best of our abilities, a swatch of space from an earlier
epoch.

Somewhere in the southern sky, pulling away from us with
the expansion of the universe, the big black hole will roll along
its own galaxy, dark and quiet until something wanders past,
an interstellar dust cloud or an errant star. After a few billion
years the host galaxy might collide with a neighbor, tossing the
black hole around, maybe toward a supermassive black hole
in a growing galactic center. Our star will die. The Milky Way
will blend with Andromeda. The record of this discovery along
with the wreckage of our solar system will eventually fall into
black holes, as will everything else in the cosmos, the expand-
ing space eventually silent, and all the black holes will evaporate into oblivion near the end of time.

ACKNOWLEDGMENTS

I am indebted to the many friends, scientists, and engineers who offered insight into the instruments and the collaboration, who took me around the labs, spent hours at a chalkboard or with a pen over paper, who told great stories, or made the trips around the country less clanging and instead very much more fun. Thank you, Rana X. Adhikari, Barry Barish, Lisa Barsotti, Aidan Brooks, Jocelyn Bell Burnell, Yanbei Chen, Ian Drever, Jenne Driggers, Matthew Evans, Joe Giaime, Peter Goldreich, Gabriela González, Eric Gustafson, Dale Ingram, Richard Isaacson, Michael Landry, Nicole Lingner, Szabi Márka, Zsuzsa Márka, Jay Marx, Nergis Mavalvala, Syd Meshkov, Brian O'Reilly, Jerry Ostriker, Larry Price, Fred Raab, Vivien Raymond, David Reitze, Jameson Rollins, Daniel Sigg, Nicolas Smith, Virginia Trimble, Tony Tyson, Robbie Vogt, Alan Weinstein, Carolee Winstein. I could have and probably should have written about all of them. Although only a few remain in the final draft, many of these contributors appeared in a once longer version of the book. Some stories were lost in the edits—with regret.

My deepest respect, admiration, and gratitude to Kip Thorne and Rainer Weiss for their generosity, for their stories, and for their time. I cannot overstate the importance to the coherence of the book of their careful, patient vetting of the history. They continually amaze me with their honesty and integrity, their cleverness and excitement and intensity. I'm honored to know them.

I am grateful to the California Institute of Technology and to the LIGO Laboratory for their hospitality. Special thanks to Sean Carroll, Carol Silberstein, and Mark Wise. My gratitude also to the Caltech archivists for their hard work and continued efforts.

This book was written in the unlikely sanctuary of two artists' studios in Brooklyn: DaMM—Dark Matter Manufacturing—and Pioneer Works. My love and gratitude to my friends there. Thank you for the essential mayhem, the crazy creative energy, the gratuitous noise, the legendary parties, and not least the crucial inspiration.

I am much obliged to Barnard College for their exceptional support over many years in general and for a Presidential Research Award in specific. I also gratefully acknowledge the essential support of a Guggenheim Fellowship during the writing of this book. Thank you, Lia Halloran and Chapman University, for hospitality during my time as a Chancellor's Fellow. Thank you, Matthew Putman, for establishing a residency at Pioneer Works and, more generally, for having and encouraging wild ideas.

Thank you, John Brockman, Katinka Matson, and Max Brockman, for getting me into this trouble. Thank you to Russell Weinberger and Warren Malone for riffing on the title.

Special gratitude to my insightful, understanding, and excellent editor, Dan Frank.

Thank you to my dear friend Pedro Ferreira for unbelievable reinforcement, always exactly when I needed to hear precisely what he was saying.

I regret that I could not represent all of the scientists and engineers crucial to the completion of the project that Kip and Rai and Ron started fifty years ago. Dozens of people deserve greater acknowledgment than I could give them. I cannot make up for the inattention, but as a proxy I can acknowledge the entire LIGO Scientific Collaboration. Following is the official LIGO author list, consisting of roughly 800 names from nearly 130 institutions from around the globe. The list includes not only the experimentalists who built the instrument, but also theorists and data analysts around the world invested in LIGO's success. Also on the list are contributors to the European Virgo project. As Rai says, "It takes a village."

THE LIGO SCIENTIFIC COLLABORATION
AND THE VIRGO COLLABORATION

B. P. Abbott, R. Abbott, T. D. Abbott, M. R. Abernathy, F. Acernese, K. Ackley, C. Adams, T. Adams, P. Addesso, R. X. Adhikari, V. B. Adya, C. Affeldt, M. Agathos, K. Agatsuma, N. Aggarwal, O. D. Aguiar, A. Ain, P. Ajith, B. Allen, A. Allocca, P. A. Altin, D. V. Amariutei, S. B. Anderson, W. G. Anderson, K. Arai, M. C. Araya, C. C. Arceneaux, J. S. Areeda, N. Arnaud, K. G. Arun, G. Ashton, M. Ast, S. M. Aston, P. Astone, P. Aufmuth, C. Aulbert, S. Babak, P. T. Baker, F. Baldaccini, G. Ballardin, S. W. Ballmer, J. C. Barayoga, S. E. Barclay, B. C. Barish, D. Barker, F. Barone, B. Barr, L. Barsotti, M. Barsuglia, D. Barta, J. Bartlett, I. Bartos, R. Bassiri, A. Basti, J. C. Batch, C. Baune, V. Bavigadda, M. Bazzan, B. Behnke, M. Bejger, C. Belczynski, A. S. Bell, C. J. Bell, B. K. Berger, J. Bergman, G. Bergmann, C. P. L. Berry, D. Bersanetti, A. Bertolini, J. Betzwieser, S. Bhagwat, R. Bhandare, I. A. Bilenko, G. Billingsley, J. Birch, R. Birney, S. Biscans, A. Bisht, M. Bitossi, C. Biwer, M. A. Bizouard, J. K. Blackburn, C. D. Blair, D. Blair, R. M. Blair, S. Bloemen, O. Bock, T. P. Bodiya, M. Boer,

G. Bogaert, C. Bogan, A. Bohe, P. Bojtos, C. Bond, F. Bondu, R. Bonnand, R. Bork, V. Boschi, S. Bose, A. Bozzi, C. Bradas-chia, P. R. Brady, V. B. Braginsky, M. Branchesi, J. E. Brau, T. Briant, A. Brillet, M. Brinkmann, V. Brisson, P. Brockill, A. F. Brooks, D. A. Brown, D. D. Brown, N. M. Brown, C. C. Buchanan, A. Buikema, T. Bulik, H. J. Bulten, A. Buon-anno, D. Buskulic, C. Buy, R. L. Byer, L. Cadonati, G. Cagnoli, C. Cahillane, J. Calderón Bustillo, T. Callister, E. Calloni, J. B. Camp, K. C. Cannon, J. Cao, C. D. Capano, E. Capocasa, F. Carbognani, S. Caride, J. Casanueva Diaz, C. Casentini, S. Caudill, M. Cavaglià, F. Cavalier, R. Cavalieri, G. Cella, C. Cepeda, L. Cerboni Baiardi, G. Cerretani, E. Cesarini, R. Chakraborty, T. Chalermsongsak, S. J. Chamberlin, M. Chan, S. Chao, P. Charlton, E. Chassande-Mottin, H. Y. Chen, Y. Chen, C. Cheng, A. Chincarini, A. Chiummo, H. S. Cho, M. Cho, J. H. Chow, N. Christensen, Q. Chu, S. Chua, S. Chung, G. Ciani, F. Clara, J. A. Clark, F. Cleva, E. Coccia, P.-F. Cohadon, A. Colla, C. G. Collette, M. Con-stancio, Jr., A. Conte, L. Conti, D. Cook, T. R. Corbitt, N. Cor-nish, A. Corsi, S. Cortese, C. A. Costa, M. W. Coughlin, S. B. Coughlin, J.-P. Coulon, S. T. Countryman, P. Couvares, D. M. Coward, M. J. Cowart, D. C. Coyne, R. Coyne, K. Craig, J. D. E. Creighton, J. Cripe, S. G. Crowder, A. Cumming, L. Cunningham, E. Cuoco, T. Dal Canton, S. L. Danilishin, S. D'Antonio, K. Danzmann, N. S. Darman, V. Dattilo, I. Dave, H. P. Daveloza, M. Davier, G. S. Davies, E. J. Daw, R. Day, D. DeBra, G. Debreczeni, J. Degallaix, M. De Laurentis, S. Deléglise, W. Del Pozzo, T. Denker, T. Dent, H. Dereli, V. Der-gachev, R. DeRosa, R. De Rosa, R. DeSalvo, S. Dhurandhar, M. C. Díaz, L. Di Fiore, M. Di Giovanni, A. Di Lieto, I. Di Palma, A. Di Virgilio, G. Dojcinoski, V. Dolique, F. Donovan, K. L. Dooley, S. Doravari, R. Douglas, T. P. Downes, M. Drago,

R. W. P. Drever, J. C. Driggers, Z. Du, M. Ducrot, S. E. Dwyer, T. B. Edo, M. C. Edwards, A. Effler, H.-B. Eggenstein, P. Ehrens, J. M. Eichholz, S. S. Eikenberry, W. Engels, R. C. Essick, T. Etzel, M. Evans, T. M. Evans, R. Everett, M. Factourovich, V. Fafone, H. Fair, S. Fairhurst, X. Fan, Q. Fang, S. Farinon, B. Farr, W. M. Farr, M. Favata, M. Fays, H. Fehrmann, M. M. Fejer, I. Ferrante, E. C. Ferreira, F. Ferrini, F. Fidecaro, I. Fiori, R. P. Fisher, R. Flaminio, M. Fletcher, J.-D. Fournier, S. Franco, S. Frasca, F. Frasconi, Z. Frei, A. Freise, R. Frey, T. T. Fricke, P. Fritschel, V. V. Frolov, P. Fulda, M. Fyffe, H. A. G. Gabbard, J. R. Gair, L. Gammaitoni, S. G. Gaonkar, F. Garufi, A. Gatto, G. Gaur, N. Gehrels, G. Gemme, B. Gendre, E. Genin, A. Gennai, J. George, L. Gergely, V. Germain, A. Ghosh, S. Ghosh, J. A. Giaime, K. D. Giardina, A. Giazotto, K. Gill, A. Glaefke, E. Goetz, R. Goetz, L. Gondan, G. González, J. M. Gonzalez Castro, A. Gopakumar, N. A. Gordon, M. L. Gorodetsky, S. E. Gossan, M. Gosselin, R. Gouaty, C. Graef, P. B. Graff, M. Granata, A. Grant, S. Gras, C. Gray, G. Greco, A. C. Green, P. Groot, H. Grote, S. Grunewald, G. M. Guidi, X. Guo, A. Gupta, M. K. Gupta, K. E. Gushwa, E. K. Gustafson, R. Gustafson, J. J. Hacker, B. R. Hall, E. D. Hall, G. Hammond, M. Haney, M. M. Hanke, J. Hanks, C. Hanna, M. D. Hannam, J. Hanson, T. Hardwick, J. Harms, G. M. Harry, I. W. Harry, M. J. Hart, M. T. Hartman, C.-J. Haster, K. Haughian, A. Heidmann, M. C. Heintze, H. Heitmann, P. Hello, G. Hemming, M. Hendry, I. S. Heng, J. Hennig, A. W. Heptonstall, M. Heurs, S. Hild, D. Hoak, K. A. Hodge, D. Hofman, S. E. Hollitt, K. Holt, D. E. Holz, P. Hopkins, D. J. Hosken, J. Hough, E. A. Houston, E. J. Howell, Y. M. Hu, S. Huang, E. A. Huerta, D. Huet, B. Hughey, S. Husa, S. H. Huttner, T. Huynh-Dinh, A. Idrisy, N. Indik, D. R. Ingram, R. Inta, H. N. Isa, J.-M. Isac, M. Isi, G. Islas, T. Isogai, B. R. Iyer, K. Izumi, T. Jacqmin,

H. Jang, K. Jani, P. Jaranowski, S. Jawahar, F. Jiménez-Forteza, W. W. Johnson, D. I. Jones, R. Jones, R. J. G. Jonker, L. Ju, H. K, C. V. Kalaghatgi, V. Kalogera, S. Kandhasamy, G. Kang, J. B. Kanner, S. Karki, M. Kasprzack, E. Katsavounidis, W. Katzman, S. Kaufer, T. Kaur, K. Kawabe, F. Kawazoe, F. Kéfélian, M. S. Kehl, D. Keitel, D. B. Kelley, W. Kells, R. Kennedy, J. S. Key, A. Khalaidovski, F. Y. Khalili, S. Khan, Z. Khan, E. A. Khazanov, N. Kijbunchoo, C. Kim, J. Kim, K. Kim, N. Kim, N. Kim, Y.-M. Kim, E. J. King, P. J. King, D. L. Kinzel, J. S. Kissel, L. Kleybolte, S. Klimenko, S. M. Koehlenbeck, K. Kokeyama, S. Koley, V. Kondrashov, A. Kontos, M. Korobko, W. Z. Korth, I. Kowalska, D. B. Kozak, V. Kringel, B. Krishnan, A. Królak, C. Krueger, G. Kuehn, P. Kumar, L. Kuo, A. Kutynia, B. D. Lackey, M. Landry, J. Lange, B. Lantz, P. D. Lasky, A. Lazzarini, C. Lazzaro, P. Leaci, S. Leavey, E. Lebigot, C. H. Lee, H. K. Lee, H. M. Lee, K. Lee, M. Leonardi, J. R. Leong, N. Leroy, N. Letendre, Y. Levin, B. M. Levine, T. G. F. Li, A. Libson, T. B. Littenberg, N. A. Lockerbie, J. Logue, A. L. Lombardi, J. E. Lord, M. Lorenzini, V. Loriette, M. Lormand, G. Losurdo, J. D. Lough, H. Lück, A. P. Lundgren, J. Luo, R. Lynch, Y. Ma, T. MacDonald, B. Machenschalk, M. MacInnis, D. M. Macleod, F. Magaña-Sandoval, R. M. Magee, M. Mageswaran, E. Majorana, I. Maksimovic, V. Malvezzi, N. Man, I. Mandel, V. Mandic, V. Mangano, G. L. Mansell, M. Manske, M. Mantovani, F. Marchesoni, F. Marion, S. Márka, Z. Márka, A. S. Markosyan, E. Maros, F. Martelli, L. Martellini, I. W. Martin, R. M. Martin, D. V. Martynov, J. N. Marx, K. Mason, A. Masserot, T. J. Massinger, M. Masso-Reid, F. Matichard, L. Matone, N. Mavalvala, N. Mazumder, G. Mazzolo, R. McCarthy, D. E. McClelland, S. McCormick, S. C. McGuire, G. McIntyre, J. McIver, D. J. McManus, S. T. McWilliams, D. Meacher, G. D. Meadors, J. Meidam, A. Melatos, G. Mendell, D. Mendoza-Gandara, R. A. Mercer, E. Merilh, M. Merzougui, S. Meshkov, C. Messenger, C. Messick, P. M. Meyers, F. Mezzani, H. Miao, C. Michel, H. Middleton, E. E. Mikhailov, L. Milano, J. Miller, M. Millhouse, Y. Minenkov, J. Ming, S. Mirshekari, C. Mishra, S. Mitra, V. P. Mitrofanov, G. Mitselmakher, R. Mittleman, A. Moggi, S. R. P. Mohapatra, M. Montani, B. C. Moore, C. J. Moore, D. Moraru, G. Moreno, S. R. Morriss, K. Mossavi, B. Mours, C. M. Mow-Lowry, C. L. Mueller, G. Mueller, A. W. Muir, A. Mukherjee, D. Mukherjee, S. Mukherjee, A. Mullavey, J. Munch, D. J. Murphy, P. G. Murray, A. Mytidis, I. Nardecchia, L. Naticchioni, R. K. Nayak, V. Necula, K. Nedkova, G. Nelemans, M. Neri, A. Neunzert, G. Newton, T. T. Nguyen, A. B. Nielsen, S. Nissanke, A. Nitz, F. Nocera, D. Nolting, M. E. N. Normandin, L. K. Nuttall, J. Oberling, E. Ochsner, J. O'Dell, E. Oelker, G. H. Ogin, J. J. Oh, S. H. Oh, F. Ohme, M. Oliver, P. Oppermann, R. J. Oram, B. O'Reilly, R. O'Shaughnessy, C. D. Ott, D. J. Ottaway, R. S. Ottens, H. Overmier, B. J. Owen, A. Pai, S. A. Pai, J. R. Palamos, O. Palashov, C. Palomba, A. Pal-Singh, H. Pan, C. Pankow, F. Pannarale, B. C. Pant, F. Paoletti, A. Paoli, M. A. Papa, H. R. Paris, W. Parker, D. Pascucci, A. Pasqualetti, R. Passaquieti, D. Passuello, Z. Patrick, B. L. Pearlstone, M. Pedraza, R. Pedurand, L. Pekowsky, A. Pele, S. Penn, R. Pereira, A. Perreca, M. Phelps, O. Piccinni, M. Pichot, F. Piergiovanni, V. Pierro, G. Pillant, L. Pinard, I. M. Pinto, M. Pitkin, R. Poggiani, A. Post, J. Powell, J. Prasad, V. Predoi, S. S. Premachandra, T. Prestegard, L. R. Price, M. Prijatelj, M. Principe, S. Privitera, R. Prix, G. A. Prodi, L. Prokhorov, M. Punturo, P. Puppo, M. Pürrer, H. Qi, J. Qin, V. Quetschke, E. A. Quintero, R. Quitzow-James, F. J. Raab, D. S. Rabeling, H. Radkins, P. Raffai, S. Raja, M. Rakhmanov, P. Rapagnani, V. Raymond, M. Razzano, V. Re, J. Read, C. M. Reed, T. Regimbau, L. Rei,

S. Reid, D. H. Reitze, H. Rew, F. Ricci, K. Riles, N. A. Robertson, R. Robie, F. Robinet, A. Rocchi, L. Rolland, J. G. Rollins, V. J. Roma, J. D. Romano, R. Romano, G. Romanov, J. H. Romie, D. Rosínska, S. Rowan, A. Rüdiger, P. Ruggi, K. Ryan, S. Sachdev, T. Sadecki, L. Sadeghian, M. Saleem, F. Salemi, A. Samajdar, L. Sammut, E. J. Sanchez, V. Sandberg, B. Sandeen, J. R. Sanders, B. Sassolas, B. S. Sathyaprakash, P. R. Saulson, O. Sauter, R. L. Savage, A. Sawadsky, P. Schale, R. Schilling, J. Schmidt, P. Schmidt, R. Schnabel, A. Schnbeck, R. M. S. Schofield, E. Schreiber, D. Schuette, B. F. Schutz, J. Scott, S. M. Scott, D. Sellers, D. Sentenac, V. Sequino, A. Sergeev, G. Serna, Y. Setyawati, A. Sevigny, D. A. Shaddock, S. Shah, M. S. Shahriar, M. Shaltev, Z. Shao, B. Shapiro, P. Shawhan, A. Sheperd, D. H. Shoemaker, D. M. Shoemaker, K. Siellez, X. Siemens, D. Sigg, A. D. Silva, D. Simakov, A. Singer, L. P. Singer, A. Singh, R. Singh, A. M. Sintes, B. J. J. Slagmolen, J. R. Smith, N. D. Smith, R. J. E. Smith, E. J. Son, B. Sorazu, F. Sorrentino, T. Souradeep, A. K. Srivastava, A. Staley, M. Steinke, J. Steinlechner, S. Steinlechner, D. Steinmeyer, B. C. Stephens, R. Stone, K. A. Strain, N. Straniero, G. Stratta, N. A. Strauss, S. Strigin, R. Sturani, A. L. Stuver, T. Z. Summerscales, L. Sun, P. J. Sutton, B. L. Swinkels, M. J. Szczepanczyk, M. Tacca, D. Talukder, D. B. Tanner, M. Tápai, S. P. Tarabrin, A. Taracchini, R. Taylor, T. Theeg, M. P. Thirugnanasambandam, E. G. Thomas, M. Thomas, P. Thomas, K. A. Thorne, K. S. Thorne, E. Thrane, S. Tiwari, V. Tiwari, K. V. Tokmakov, C. Tomlinson, M. Tonelli, C. V. Torres, C. I. Torrie, D. Töyrä, F. Travasso, G. Traylor, D. Trifiro, M. C. Tringali, L. Trozzo, M. Tse, M. Turconi, D. Tuyenbayev, D. Ugolini, C. S. Unnikrishnan, A. L. Urban, S. A. Usman, H. Vahlbruch, G. Vajente, G. Valdes, N. van Bakel, M. van Beuzekom, J. F. J. van den Brand, C. van den Broeck, L. van der Schaaf, M. V. van der Sluys, J. V. van Heijningen, A. A. van Veggel, M. Vardaro, S. Vass, M. Vasúth, R. Vaulin, A. Vecchio, G. Vedovato, J. Veitch, P. J. Veitch, K. Venkateswara, D. Verkindt, F. Vetrano, A. Viceré, S. Vinciguerra, J.-Y. Vinet, S. Vitale, T. Vo, H. Vocca, C. Vorvick, W. D. Vousden, S. P. Vyatchanin, A. R. Wade, L. E. Wade, M. Wade, M. Walker, L. Wallace, S. Walsh, G. Wang, H. Wang, M. Wang, X. Wang, Y. Wang, R. L. Ward, J. Warner, M. Was, B. Weaver, L.-W. Wei, M. Weinert, A. J. Weinstein, R. Weiss, T. Welborn, L. Wen, P. Wessels, T. Westphal, K. Wette, J. T. Whelan, S. E. Whitcomb, D. J. White, B. F. Whiting, R. D. Williams, A. R. Williamson, J. L. Willis, B. Willke, M. H. Wimmer, W. Winkler, C. C. Wipf, H. Wittel, G. Woan, J. Worden, J. L. Wright, G. Wu, J. Yablon, W. Yam, H. Yamamoto, C. C. Yancey, M. J. Yap, H. Yu, M. Yvert, A. Zadroźny, L. Zangrando, M. Zanolin, J.-P. Zendri, M. Zevin, F. Zhang, L. Zhang, M. Zhang, Y. Zhang, C. Zhao, M. Zhou, Z. Zhou, X. J. Zhu, M. E. Zucker, S. E. Zuraw, J. Zweizig.

NOTES ON SOURCES

1. WHEN BLACK HOLES COLLIDE

The quotation on page 5, "a change in distance comparable to less than a human hair relative to 100 billion times the circumference of the world," is excerpted from the following source: Tyson, Anthony. Testimony for the House of Representatives' hearing of the Committee on Science, Space, and Technology, March 13, 1991.

2. HIGH FIDELITY

Throughout the book, I edited together several of my own interviews with Rainer Weiss conducted over many meetings between 2013 and 2015 with the interview by Shirley Cohen for the Caltech Archives Oral History Project, cited below. In some cases, Rai's responses were similar enough to those in the Shirley Cohen interview that I preferred to use her transcription out of deference to the earlier date of her interview.

Weiss, Rainer. Interview by Shirley Cohen. Pasadena, California, May 10, 2000. Oral History Project, California Institute of Technology Archives.

Weiss, Rainer. Interviews by the author. From a series of interviews between 2013 and 2015.

3. NATURAL RESOURCES

All quotes from Kip Thorne are drawn from Thorne, Kip. Interview by the author. From a series of interviews between 2013 and 2015.

For much of the history of general relativity I relied on the excellent book by the astrophysicist Pedro Ferreira: *A Perfect Theory: A Century of Geniuses and the Battle over General Relativity*. New York: Houghton Mifflin Harcourt Publishing Company, 2014.

Thorne, Kip S. *Black Holes and Time Warps: Einstein's Outrageous Legacy*. New York: W. W. Norton & Company, Inc., 1995.

Wheeler, John Archibald. *Geons, Black Holes, and Quantum Foam: A Life in Physics*. New York: W. W. Norton & Company, Inc., 1998.

The number of PhDs Wheeler mentored was taken from Terry M. Christensen, "John Wheeler's Mentorship: An Enduring Legacy," *Physics Today* 62, no. 4 (August 2009): 55.

4. CULTURE SHOCK

Drever, Ronald P. Interview by Shirley Cohen. Pasadena, California, May 10, 2000. Oral History Project, California Institute of Technology Archives.

Outside of Ron's own words, I am grateful to Ian Drever for sharing his recollections of their childhood. I borrow liberally and gratefully from the following document: Drever, John (Ian). Private communication of a document Dr. Ian Drever wrote about his family and his older brother Ronald, October 2015.

5. JOE WEBER

All quotes from Joe Weber are drawn from Weber, Joseph. Interview by Kip Thorne conducted during research on his book *Black Holes and Time Warps: Einstein's Outrageous Legacy,* July 20, 1982. California Institute of Technology Archives.

All quotes from Ron Drever drawn from Drever, Ronald P. Interview by Shirley Cohen. Pasadena, California. Session 1: January 21, 1997. Session 2: February 10, 1997. Session 3: February 25, 1997. Session 4: March 13, 1997. Session 5: June 3, 1997. Oral History Project, California Institute of Technology Archives.

Bartusiak, Marcia. *Einstein's Unfinished Symphony: Listening to the Sounds of Spacetime,* Washington, D.C.: Joseph Henry Press, 2000.

Collins, Harry. *Gravity's Shadow: The Search for Gravitational Waves.* Chicago: The University of Chicago Press, 2004.

Dyson, Freeman. "Gravitational Machines." In *Interstellar Communication: A Collection of Reprints and Original Contributions,* edited by A. G. W. Cameron, 115. New York: W. A. Benjamin, 1963.

All quotes from Tony Tyson are drawn from Tyson, Anthony. Interview by the author, 2015.

6. PROTOTYPES

Weiss, Rainer. Interviews by the author. From a series of interviews between 2013 and 2015.

Weiss, Rainer. Interview by Shirley Cohen. Pasadena, California, May 10, 2000. Oral History Project, California Institute of Technology Archives.

Mikhail E. Gertsenshtein and V. I. Pustovoit, "On the Detection of Low-Frequency Gravitational Waves," *Soviet Physics—JETP* 16, (1963): 433–435.

All quotes from Kip Thorne are drawn from Thorne, Kip. Interviews by the author. From a series of interviews between 2013 and 2015.

Isaacson, Richard. Interview by the author, 2015.

7. THE TROIKA

Weiss, Rainer. Interviews by the author. From a series of interviews between 2013 and 2015.

Weiss, Rainer. Interview by Shirley Cohen. Pasadena, California, May 10, 2000. Oral History Project, California Institute of Technology Archives.

All quotes from Ron Drever are drawn from Drever, Ronald P. Interview by Shirley Cohen. Pasadena, California. Session 1: January 21, 1997. Session 2: February 10, 1997. Session 3: February 25, 1997. Session 4: March 13, 1997. Session 5: June 3, 1997. Oral History Project, California Institute of Technology Archives.

8. THE CLIMB

All quotes from Ron Drever are drawn from Drever, Ronald P. Interview by Shirley Cohen. Pasadena, California. Session 1: January 21, 1997. Session 2: February 10, 1997. Session 3: February 25, 1997. Session 4: March 13, 1997. Session 5: June 3, 1997. Oral History Project, California Institute of Technology Archives.

Bell Burnell, Jocelyn. Interview by the author, 2015.

The quotation on page 96, "Miss Bell, you have made . . . ," is excerpted from Longair, Malcolm. *The Cosmic Century: A History of Astrophysics and Cosmology.* Cambridge, UK: Cambridge University Press, 2006.

9. WEBER AND TRIMBLE

All quotes from Virginia Trimble are drawn from Trimble, Virginia. Interview by the author, 2014, except those specifically referenced in "Behind a Lovely Face, a 180 I.Q." *Life*, October 19, 1962, pp. 98–99.

All quotes from Joe Weber are drawn from Weber, Joseph. Interview by Kip Thorne conducted during research on his book *Black*

Holes and Time Warps: Einstein's Outrageous Legacy, July 20, 1982. California Institute of Technology Archives.

Freeman Dyson letter from Collins: Collins, Harry. *Gravity's Shadow: The Search for Gravitational Waves.* Chicago: The University of Chicago Press, 2004.

10. LHO

Landry, Michael. Interviews by the author. From a series of interviews between 2012 and 2015.

Weiss, Rainer. Interviews by the author. From a series of interviews between 2013 and 2015.

Weiss, Rainer. Interview by Shirley Cohen. Pasadena, California, May 10, 2000. Oral History Project, California Institute of Technology Archives.

11. SKUNKWORKS

The quote on page 132 "If I went back in, my colleagues would have . . ." is excerpted from a Caltech article to be found at http://calteches.library.caltech.edu/3432/1/Vogt.pdf.

Other quotes from Robbie Vogt are from Vogt, Rochus. Interview by the author, 2014.

Collins, Harry. *Gravity's Shadow: The Search for Gravitational Waves.* Chicago: The University of Chicago Press, 2004.

Quotes from Tony Tyson are from Tyson, Anthony. Interview by the author, 2015.

12. GAMBLING

Ostriker, Jeremiah. Interview by the author, 2015.

Thorne, Kip. Interview by the author. From a series of interviews between 2013 and 2015.

The Stephen Hawking quote is from Thorne, Kip S. *Black Holes and Time Warps: Einstein's Outrageous Legacy.* New York: W. W. Norton & Company, Inc., 1995.

See also: Hawking, Stephen. *A Brief History of Time.* New York: Bantam Dell Publishing Group, 1988.

13. RASHOMON

All quotes from Robbie Vogt are from Vogt, Rochus. Interview by the author, 2014.

All quotes from Stan Whitcomb are from Whitcomb, Stanley. Interviews by the author. From a series of interviews between 2012 and 2015.

All quotes from Ron Drever drawn from Drever, Ronald P. Interview by Shirley Cohen. Pasadena, California. Session 1: January 21, 1997. Session 2: February 10, 1997. Session 3: February 25, 1997. Session 4: March 13, 1997. Session 5: June 3, 1997. Oral History Project, California Institute of Technology Archives.

Weiss, Rainer. Interviews by the author. From a series of interviews between 2013 and 2015.

Weiss, Rainer. Interview by Shirley Cohen. Pasadena, California, May 10, 2000. Oral History Project, California Institute of Technology Archives.

Goldreich, Peter. Interview by Shirley Cohen. Pasadena, California, March, April, November 1998. Oral History Project, California Institute of Technology Archives.

The memo referenced on p. 163 is undoubtedly in a manuscript collection that appears as an entry in the Caltech Archives titled Documents of the Drever-LIGO Controversy. Those documents are still sealed at the time of this writing.

14. LLO

Braginsky, Vladimir. Interview by Shirley Cohen, Pasadena, California, January 15, 1997. Oral History Project, California Institute of Technology Archives.

Adhikari, Rana X. Interviews by the author. From a series of interviews between 2011 and 2015.

O'Reilly, Brian. Interviews by the author. From a series of interviews between 2013 and 2015.

González, Gabriela. Interviews by the author. From a series of interviews between 2013 and 2015.

Giaime, Joe. Interviews by the author. From a series of interviews between 2013 and 2015.

Barish, Barry. Interviews by the author. From a series of interviews between 2013 and 2015.

16. THE RACE IS ON

Vogt, Rochus. Interview by the author, 2014.

Thorne, Kip. Interviews by the author. From a series of interviews between 2013 and 2015.

Weiss, Rainer. Interviews by the author. From a series of interviews between 2013 and 2015.

Hough, James. Interview by the author, 2015.

INDEX

A Note About the Author

Janna Levin is a professor of physics and astronomy at Barnard College of Columbia University and a recent Guggenheim fellow. Her scientific research concerns the early universe and black holes. Her last book, *A Madman Dreams of Turing Machines,* won the PEN/Bingham Prize. She is also the author of *How the Universe Got Its Spots: Diary of a Finite Time in a Finite Space.*

A Note on the Type

The text of this book was set in Sabon, a typeface designed in 1966 in Frankfurt by Jan Tschichold (1902–1974), the well-known German typographer, and named for the famous Lyons punch cutter Jacques Sabon. Based loosely on the original designs by Claude Garamond (ca. 1480–1561), Sabon is unique in that it was explicitly designed for hot-metal composition on both the Monotype and Linotype machines as well as for filmsetting.

Typeset by Scribe, Philadelphia, Pennsylvania

Printed and bound by R.R. Donnelley,
Harrisonburg, Virginia

Designed by Betty Lew